Assessing
Hands-On
Science

**CORWIN
PRESS**

The Corwin Press logo—a raven striding across an open book—represents the happy union of courage and learning. We are a professional-level publisher of books and journals for K-12 educators, and we are committed to creating and providing resources that embody these qualities. Corwin's motto is "Success for All Learners."

Assessing Hands-On Science

A Teacher's Guide to Performance Assessment

Janet Harley Brown
Richard J. Shavelson

CORWIN PRESS, INC.
A Sage Publications Company
Thousand Oaks, California

For information address:

Corwin Press, Inc.
A Sage Publications Company
2455 Teller Road
Thousand Oaks, California 91320
E-mail: order@corwin.sagepub.com

SAGE Publications Ltd.
6 Bonhill Street
London EC2A 4PU
United Kingdom

SAGE Publications India Pvt. Ltd.
M-32 Market
Greater Kailash I
New Delhi 110 048 India

Printed in the United States of America

Library of Congress Cataloging-in-Publication Data

Brown, Janet Harley.
    Assessing hands-on science: a teacher's guide to performance
assessment / authors, Janet Harley Brown and Richard J. Shavelson.
        p. cm.
    Includes bibliographical references (p.).
    ISBN 0-8039-6442-0 (cloth: acid-free paper). —
    ISBN 0-8039-6443-9 (pbk.: acid-free paper).
    1. Science—Ability testing. 2. Educational tests and
measurements. I. Shavelson, Richard J. II. Title.
LB1585.B687 1996
372.3'5'044—dc20   96-4473

96 97 98 99 10 9 8 7 6 5 4 3 2 1

Corwin Production Editor: Diana E. Axelsen
Corwin Typesetter: Janelle LeMaster
Cover Illustration: Willis Copeland

Contents

Preface

Assessing "performance" in the classroom makes a lot of intuitive sense to teachers. We watch students perform as they tackle a task and we instinctively judge that performance. But how does that casual, but constant, observation by the teacher differ from "performance assessment"? This book offers teachers and other educators an opportunity to learn about this new testing technology from the inside out—that is, by actual hands-on experience with a variety of science performance assessments.

Why Use Performance Assessments?

The current reform in education is based, in part, on the idea that for students to learn, they must *construct* knowledge for themselves—not merely memorize facts. To this end, the movement toward hands-on science in elementary and middle school classrooms rests on students *doing* science, not just learning *about* it from teachers and books. Students construct knowledge as they dig in and investigate the world around them.

All too often, however, students are tested primarily on what they have memorized. The tests seem only partially relevant to students and, at the same time, they give students a very narrow message about what is important to learn in science. Performance assessments offer students and teachers a way to assess knowledge and procedural skills that is consistent with a hands-on curriculum. The assessments look and feel like the learning activities students engage in, but they are standardized to meet rigorous testing criteria. When used in conjunction with well-constructed paper-and-pencil tests, they provide teachers with rich information about the conceptual and procedural knowledge of students.

The Purpose of This Book

We have found that many teachers, and other educators, have questions about performance tests. What are they like? What do they really measure? Most important, what qualities should they have to make them trustworthy testing instruments? Performance tests look like a lot of extra work to learn to use—so are they really worth it? Are they all they're cracked up to be?

For the past 5 years, our research team has been addressing these questions. We have developed and analyzed performance assessments for specific hands-on science curricula and have developed a teacher enhancement workshop to share what we have learned about performance assessment with the educational community. This workshop has been given in Arizona, Alaska, California, and Washington, D.C.

This book takes the activities and information from the workshop and allows you, the reader, to learn about this new testing technology at your own pace, constructing knowledge—hands-on. The book, therefore, is essentially a workbook, with assessments to try for yourself, scoring systems to learn about, and implications of scores to consider for instruction and accountability purposes. We introduce you to several different types of assessments and different types of score forms. Also, because performance assessments are very new in elementary and middle school science, we include a chapter devoted to helping you recognize what constitutes a good performance assessment—one you can trust to provide reliable, valid, and useful information about what your students know and can do in science.

This book is intended for teachers who currently use (or are planning to use) a hands-on science curriculum. In addition, it would be useful to preservice teachers, graduate students in education, and other educators interested in alternative forms of assessment.

Acknowledgments

All the materials in this book are the culmination of the efforts of many people—researchers, teachers, scientists, and other educators. We gratefully acknowledge the members of our research team for their hard work and support: Gail Baxter, Willis Copeland, Lynn Decker, Steve Druker, Maria Araceli Ruiz-Primo, and Guillermo Solano-Flores. Also, many thanks to Marilyn Bachmann, Xiaohong Gao, Abdul Rahman Othman, Jerry Pine, Laurie Thompson, Jennifer Yuré, and the many teachers we worked with in the Pasadena, California, Unified School District. We also thank the National Science Foundation for funding our research and development work, especially Susan Snyder, whose support and encouragement are very much appreciated.

About the Authors

Janet Harley Brown is a former elementary school teacher who has been involved in educational research for the past 8 years. Her current work has involved developing and presenting teacher enhancement workshops in "Knowledge, Use, and Selection of Performance Assessments in Science." She received her M.A. and Ph.D. in educational psychology from the University of California, Santa Barbara, where she is currently a Researcher/Lecturer.

Richard J. Shavelson is the I. James Quillen Dean of the School of Education and Professor of Education and Psychology at Stanford University. He is former Dean of the Graduate School of Education at the University of California, Santa Barbara; former president of the American Educational Research Association; and chair of the National Academy of Science's Board on Testing and Assessment. His current research focuses on testing reform and ranges from developing psychometric models for performance assessments to developing enhancement programs for teachers in using new forms of tests to analyzing the policy implications of the testing reform.

Why Use
Performance Assessment?

Overview

Perhaps you, like other teachers across the country, have felt that there is a definite mismatch between your hands-on, inquiry-based science curriculum and old traditional paper-and-pencil methods for testing student knowledge. You have heard about performance assessment and are wondering if it really can provide a more complete picture of what your students know and can do in science.

This chapter introduces you to performance assessment by asking, point-blank, *Why?*—Why give performance assessments?

The Issue

In a fifth-grade classroom in Santa Barbara, students, working in pairs, are busily doing chemical tests to distinguish among a group of common household powders. The students are not only learning about the physical and

1

chemical properties of the powders before them, they are also learning to observe, gather evidence to identify a particular powder, and draw conclusions based on that evidence. Their teacher, Marilyn, moves from pair to pair, reminding students that tests can be used either to confirm the identity of a powder or to disconfirm the presence of another powder, thereby indicating what the powder is not. Students are learning how to think scientifically and do science. As the lesson comes to a close, Marilyn wonders if her students have really caught on to the concepts and procedures presented so far in the unit. How should she go about testing student knowledge? In fact, Marilyn had been wrestling with this question for years.

A Curriculum/Testing Mismatch

Very early in her teaching career, Marilyn caught on to the advantages of having students *do* science rather than just talk about it or read about it. But although her students loved the science labs, they hated her paper-and-pencil tests—tests that invariably included some multiple-choice questions, a few fill-in-the-blank, and a short-answer question or two. She told us recently, "The students thought these tests were 'unfair' because they were nothing like what we were doing in the lab."

When Marilyn found that most of her students didn't do well on the paper-and-pencil tests, she would try to compensate by spending the last two class sessions of each unit "teaching the test." She continued,

> I would drill on vocabulary and facts, and as a result, my students would do better, but I knew we wasted valuable lab days on drill and, in the end, I had no idea if students really understood the scientific concepts and procedures they had worked with. The tests simply didn't assess all of what I was trying to teach through the labs.

Why Performance Assessments?

Does Marilyn's dilemma sound familiar? If it does, performance assessment could be just what you are looking for. It was for Marilyn. She says,

> Why I didn't think of performance assessment myself I'll never know, but when I discovered it the relief was tremendous. In fact, as a result my life as a science teacher has changed dramatically. With a good performance assessment, students are asked to solve significant scientific problems like those they encounter in everyday life. They may be asked to design an experiment, conduct it, and then analyze the results. They may also be asked to apply what they have learned in the lab to solve a new problem. In other words, performance assessments ask students to *be* scientists and to *use* scientific procedures to solve real problems. They ask students to *think*.

(Incidentally, if you are still wondering about how Marilyn should test her students on their investigation of powders, hang on—Chapter 5 is entirely devoted to assessing a classroom unit, using Mystery Powders as the example.)

Reform Is in the Air

Science teaching is changing and this change is based, for the most part, on our conception of how learning occurs in the first place. Students integrate new knowledge with previously held knowledge as they make sense of the world around them and, in so doing, they *construct* knowledge for themselves. In the classroom, rote memorization of scientific facts has given way to direct experience with scientific phenomena—hands-on activities—that facilitate the construction of knowledge. Students do science, not just learn about it.

So, how does this reform actually play out in the classroom? Following is a brief description of some of the major differences in science instruction before and during the current reform.

Characteristics of Science Curriculum

Pre-Reform	Reform
Emphasis on knowing	Emphasis on doing as well as knowing
Students' primary learning tools are the textbook and the notepad	Students' primary learning tools are lab notebooks and manipulatives
Broad coverage of many topics	In-depth coverage of a few topics
Students work individually	Students may work individually or with partners
Spectator-based format	Activity-based format

Before the reform, science instruction was primarily text driven. Students were presented with the facts and principles of science by reading about them and then (sometimes) participated in labs or watched the teacher do demonstrations that illustrated those facts and principles. Now students learn by doing—they *discover* scientific principles by working with the real stuff—rocks, magnets, inclined planes, circuits, plants, animals—learning to think scientifically as they inquire about the world around them. Often placing students in groups, teachers concentrate on teaching a few topics in depth rather than more superficially covering many topics.

Unfortunately, most current *assessment* practices do not reflect this shift from rote learning to construction of knowledge through hands-on activities. In fact, science teachers all over the country have been struggling with the lack of congruence between the new ways of teaching and the old ways of testing.

Matching Assessment With Curriculum

As Marilyn found in her classroom, performance assessment is like the other side of the coin to a hands-on science curriculum—the two complement each other. Unlike paper-and-pencil tests, these assessments look and feel like the learning activities students do in their daily labs. But the assessments differ from the labs in that they have been *standardized* through rigorous development to ensure that they are trustworthy measurement tools. They must be able to be scored consistently by different scorers and must truly be testing the concepts and procedures we think they are testing. (The qualities that make a good assessment are discussed at length in Chapter 7.)

The following diagram shows how testing must change to be compatible with the reform in curriculum. We are assessing not only *what* students know (paper-and-pencil tests may be appropriate measures of some of this knowledge) but also if they know *how* to apply that knowledge to solve problems (performance assessments may be appropriate measures of this knowledge in action). They are evaluated on the basis of how they went about solving the problem, not just whether they arrived at a correct answer. Furthermore, they are given concrete materials to work with—individually, with a partner, or in groups. The focus with this type of assessment is on depth of understanding—not recalling a myriad of facts.

Characteristics of Science Assessment

Pre-Reform	Reform
Emphasis on assessment of knowledge	Assessment of procedural and conceptual knowledge
Uses paper and pencil	Uses manipulatives
Broad assessment of many topics	In-depth assessment of a few topics
Students assessed individually	Students may be assessed while working in groups
Focus is on correct answers	Focus is on reasonableness of procedures and correct answers

Although knowing the facts and concepts of science is crucial, because students need this knowledge to do science, determining whether they have memorized facts and concepts is not enough. Students need to be able to demonstrate that they can apply their knowledge in real, hands-on situations. So, in the assessment of learning in hands-on science, the teacher's central question is, "How do I as a teacher know that my students can do science?" Good performance assessments are designed to help you answer this question.

Summary

Why do we need to use performance assessments in our science classrooms? The answer lies in examining the way our teaching of science has changed from a textbook-driven approach to a hands-on approach. Performance assessments are simply the flip side of the coin to an inquiry-based approach to science teaching, allowing us to assess whether our students can do science and not just know about it.

What Do Performance
Assessments Measure?

What does a performance assessment look like, feel like, and most important of all, what information does it give you about what students know and can do in science? In this chapter, you will experience firsthand a performance assessment, seeing for yourself what it's like and what it measures. But first we'll refresh your memory about that familiar old testing format, the multiple-choice test, so we can compare it to performance assessment. While you are taking the two kinds of tests, we will ask you to reflect on some questions designed to probe important differences between multiple-choice tests and performance assessments.

You could just sit there in your easy chair and read about performance assessments, but if you really want to know what they are all about you need to do them—*constructing* your knowledge about performance assessments for yourself.

How are you going to manage that, you ask? Well, at the beginning of most chapters, you will find a list of materials you need to do the assessments in that chapter. We have purposely chosen assessments with materials that are commonly available and relatively easy to assemble. So as you progress through this book, we highly recommend that you *do* all the assessments and exercises for yourself—real hands-on stuff—not just imagine yourself doing them. Thinking about them in the abstract will not begin to inform you as much as it will to dig in and grapple with each task for yourself.

We suggest you gather all the materials for each chapter ahead of time and have them ready to use as you proceed through the chapter. Here are the materials you will need for Chapter 2:

HOUSE

**Performance Assessment
(4th/5th grade)**

1. 2 D-size batteries

2. 4 pieces of wire

3. 2 small flashlight bulbs with holders that allow you to connect the bulbs to the wires

4. Picture of House—removed from Resource A

Remembering the Multiple-Choice Test

Before you try a performance assessment, we want to give you some standard for comparison—typical multiple-choice questions like those commonly used to test student knowledge of electricity. As you take the test on page 8, ask yourself these questions:

1. What kinds of knowledge did it take for me to answer the questions?
2. How does the test make me feel?
3. Would I be comfortable having my achievement in science measured by a test like this?
4. As a teacher, how would I teach if I knew my students were going to be evaluated on a test like this?

Imagine a test administrator standing before you. He says, "Follow these directions exactly":

- You have 3 minutes to complete the test.
- You are to work alone. . . . No talking.
- If you are not sure of an answer but can reduce some of the alternatives, you may guess at the answer. You may begin.

Multiple-Choice Science Test

1. Choose the correct answer(s). (You may choose more than one.)

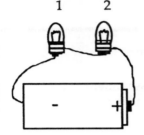

   A. Bulb 1 is brighter than bulb 2.
   B. Bulb 2 is brighter than bulb 1.
   C. Bulb 1 is not lit.
   D. Bulb 2 is not lit.
   E. Bulbs 1 and 2 are the same
       brightness.

2. Which of the following is an example of reverse polarity?

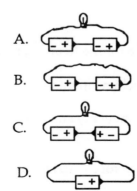

3. Which will make the bulb(s) light?

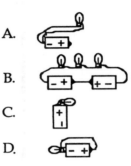

4. The unit used to measure the amount of current in a circuit is:

   A.   volts
   B.   ohms
   C.   resistance
   D.   amps

Figure 2.1. Multiple-Choice Science Test
Brown and Shavelson, *Assessing Hands-On Science: A Teacher's Guide to Performance Assessment.* Copyright 1996, Corwin Press, Inc.

Think and Respond

Now consider the following four questions. Jot down your responses and then read what other teachers have said when asked to respond to the same questions.

1. What kinds of knowledge did it take for you to answer the multiple-choice questions?

_____

_____

_____

2. How does the test make you feel?

_____

_____

_____

3. Would you be comfortable having your achievement in science measured by a test like this?

_____

_____

_____

4. As a teacher, how would you teach if you knew your students were going to be evaluated on a test like this?

_____

_____

_____

Here are some typical responses teachers have given to these questions at our workshops:

1. What kinds of knowledge did it take for you to answer the multiple-choice questions?

- Questions 2 and 4 called for remembering a vocabulary term and its underlying meaning.
- The questions were strictly recall and if you hadn't memorized the particular fact, you just guessed.

2. How does the test make you feel?

- Nervous. Some of the facts don't seem important enough to remember, and a simple mistake or misreading of what the test developer wants could negatively affect how well I do.
- The test is unnatural to doing science. Science does not present a set of alternatives to choose from. Doing science often means that the materials are concrete and react to what you do to them—not so with paper-and-pencil tests.

3. Would you be comfortable having your achievement in science measured by a test like this?

- There is more to science than just facts or giving a label to things.
- Science involves solving problems based on what you know and think might be so. It involves concrete tests of ideas—the gathering of evidence to support conclusions.

4. As a teacher, how would you teach if you knew your students were going to be evaluated on a test like this?

- I would concentrate on making sure students memorized facts and concepts.
- I would teach test-taking skills—like eliminating alternatives on multiple-choice tests.

Components of a Multiple-Choice Test

Before we move on to considering performance assessments, think for a moment about the component parts of a multiple-choice test. You are presented with a *question or problem* whose context is strictly limited to a few words. Then you *respond* by selecting one correct answer from among several other incorrect alternatives. You find out how you did after someone (probably your teacher) uses a *scoring* key listing all the correct answers. Your score is the total or percentage of correct answers.

Here is the scoring key for the test you just took. As you score your test, reflect on just what your score tells you about your knowledge of electricity.

1. E; 2. C; 3. D; 4. D

Experiencing a Performance Assessment

Now we want you to use those materials you gathered to try a performance assessment. (See the list on page 7.) We will be asking you to consider the same questions you just answered, so keep them in mind. Here they are again:

1. What kinds of knowledge did it take for me to answer the questions?
2. How does the test make me feel?

3. Would I be comfortable having my achievement in science measured by a test like this?
4. As a teacher, how would I teach if I knew my students were going to be evaluated on a test like this?

The assessment you will do is House; you will need the two batteries, two bulbs in bulb holders, four wires, and the picture of the House found in Resource A.

House was designed to be given after the eighth lesson in a fourth- /fifth-grade unit called "Circuits and Pathways" (see Resource D for information about ordering this unit). By this lesson, students have learned how to construct a circuit so they can light a bulb with a battery and wires. Also, they have made circuits with more than one battery and bulb and have had hands-on experience with how different combinations of batteries and bulbs in a circuit affect bulb brightness.

Go ahead and do the assessment on pages 12 through 14 and then return to this page.

Think and Respond

Now that you have experienced a performance assessment, respond to the questions again.

1. What kinds of knowledge did it take for you to solve the problems?

_____

_____

_____

2. How does the test make you feel?

_____

_____

_____

3. Would you be comfortable having your achievement in science measured by a test like this?

_____

_____

_____

4. As a teacher, how would you teach if you knew your students were going to be evaluated on a test like this?

_____

_____

_____

HOUSE

Name _____   Partner's Name _____

Problem

Look at the house to the left.
You and your partner are in the same house and want to use the light outside the house to tell your friends to come for a visit or to call you.

Instructions:

Make two different circuits.

Circuit 1:   Use any of the materials in front of you to make a circuit with a **dim light** to tell your friends to visit.

Circuit 2:   Use any of the materials in front of you to make a circuit with a **bright light** to tell your friends to call.

NOTE: If the bulb is off it means you are not at home.

NOTE: You cannot turn the bulb on and off to send your friends messages. The signal must work even if you cannot be near it.

Go ahead and build your circuits. You may use the imaginary house provided. You may use page 2 to keep notes on the things you do and observations you make. When you are finished making your circuits draw your answers in the spaces on page 3.

Figure 2.2. (1 of 3) House Performance Assessment

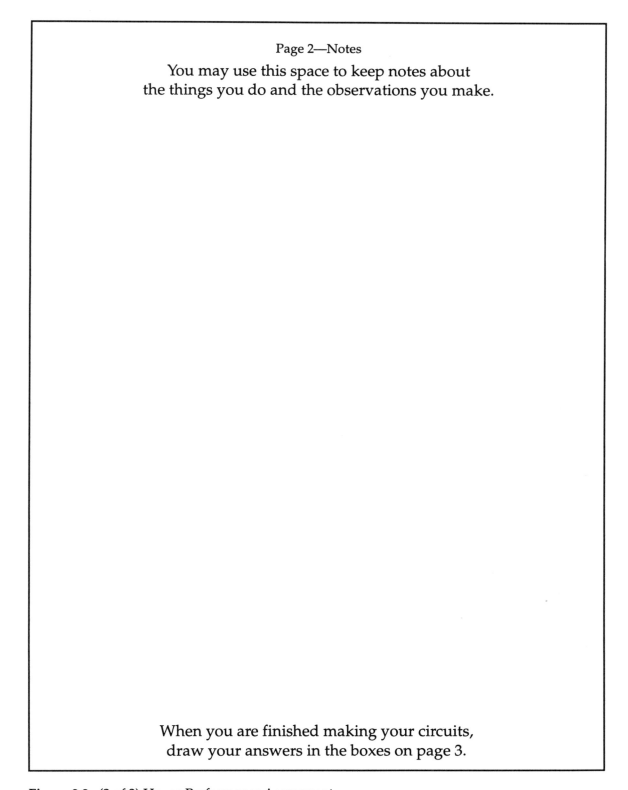

Page 2—Notes
You may use this space to keep notes about
the things you do and the observations you make.

When you are finished making your circuits,
draw your answers in the boxes on page 3.

Figure 2.2. (2 of 3) House Performance Assessment
Brown and Shavelson, *Assessing Hands-On Science: A Teacher's Guide to Performance Assessment.* Copyright 1996, Corwin Press, Inc.

Visit (dim)

Draw the **dimmer** circuit you made for **visit** in this box

Call (bright)

Draw the **brighter** circuit you made for **call** in this box

Figure 2.2. (3 of 3) House Performance Assessment

Brown and Shavelson, *Assessing Hands-On Science: A Teacher's Guide to Performance Assessment.* Copyright 1996, Corwin Press, Inc.

Here are some of the comments teachers have made in response to these questions after doing the performance assessment you just did.

1. What kinds of knowledge did it take for you to solve the problems?

    ■ Facts and concepts have to be applied in solving a concrete problem, and the solution can be tested right then and there. For example, if I know that I can vary light brightness by different configurations of circuits, I can test this out to solve the House problem.
    ■ Unlike the multiple-choice test, the assessment uses science process skills, such as observing, comparing, communicating, and relating concrete and abstract ideas.

2. How does the test make you feel?

    ■ This is fun.
    ■ I am free to investigate as I want to. The materials react to what I do and I can adjust my performance accordingly.
    ■ This doesn't seem like a test.

3. Would you be comfortable having your achievement in science measured by a test like this?

    ■ Yes. This provides an opportunity to show what I can do in a relatively realistic, concrete situation closely linked to the curriculum.

4. As a teacher, how would you teach if you knew your students were going to be evaluated on a test like this?

    ■ I would increase use of hands-on science activities and investigations, encouraging students to discover and construct knowledge.
    ■ I would attempt to find assessments like this for my science units.

Components of a Performance Assessment

Just as a multiple-choice test can be divided into its component parts (*question* or problem, selected *response*, and *scoring* key listing correct answers), a performance assessment can also be divided into components. A performance assessment is made up of a *task* or problem to solve, a format in which to *respond*, and a *scoring* system. You have experienced both the task and the response format for the preceding assessment, but what about the scoring system?

How do you think your performance on this assessment should be evaluated? If you were designing a scoring system for House, on what aspects of the student's performance would you focus? Spend a few minutes thinking about the knowledge that is demonstrated through successful completion of the task.

Jot down some notes below on how you could set up a scale to judge the quality of the various solution paths.

_____

_____

_____

_____

Here is a very simple score form that was developed for the House Assessment (Figure 2.3). You will see that, unlike the multiple-choice test, not only a correct answer is scored but also the way that answer was reached. Performance is evaluated on three aspects. First of all, we consider whether the student accomplishes the goal of constructing circuits with varying bulb brightnesses. By successfully completing this part, the student demonstrates the ability to construct a complete circuit and has made two circuits that vary in bulb brightness—so 1 point is given here. Next we see whether the circuits constructed were systematic and efficient. In other words, did the student make the dim signal by using one *battery* and the bright signal by using two *batteries* (or the dim with two *bulbs* and the bright with one *bulb*) when creating the differences in bulb brightness. Students had been learning about series circuits and bulb resistance in lessons before this assessment, so we judge their

HOUSE ASSESSMENT SCORE FORM

DATE: ___ / ___ / ___   STUDENT: _____

Makes two circuits with different brightnesses.	☐	1

and

Produces different brightnesses by varying either the number of bulbs or the number of batteries.	☐	1

and

Circuits are neat (no batteries with opposite polarities, no crossed wires).	☐	1

Total Score
(Max. 3)   ☐

Figure 2.3. House Score Form
Brown and Shavelson, *Assessing Hands-On Science: A Teacher's Guide to Performance Assessment.*
Copyright 1996, Corwin Press, Inc.

performance on the basis of how well they applied this knowledge—and that is worth 1 point. Then we look at the drawings the student did (or their descriptions in words), giving 1 point if the circuits drawn or described are feasible and accurately depicted. This item evaluates whether the brightnesses produced are the result of student's systematic actions rather than chance. The student can earn up to 3 points on this assessment.

Comparing Multiple-Choice and Performance Assessments

Another way to compare multiple-choice tests and performance assessments is to focus on the component parts of each as summarized below. Notice that while the task on a multiple-choice test is *decontextualized and abstract*, the task of a performance assessment offers the students a *concrete, contextualized* problem and gives them concrete materials with which to solve the problem. Students are asked to *select* a response on a multiple-choice test, but they *construct* a response on a performance assessment. Finally, each item on a multiple-choice test is scored *right/wrong*, but a performance assessment judges both *answers* and the reasonableness of the *procedures* that led to those answers.

Comparing the Multiple-Choice Test with Performance Assessment

Component	Multiple-Choice	Performance Assessment
Task	Abstract Decontextualized	Concrete Contextualized
Response Format	Selected	Constructed
Scoring System	Right/Wrong	Evaluates procedures as well as answers

Our comparison between multiple-choice items and performance assessments is not intended to imply that there is not a place for multiple-choice tests in our testing programs. Well-constructed multiple-choice items can be extremely useful and efficient in assessing students' ability to both remember facts and concepts and think through abstract problems. In addition to paper-and-pencil tests, however, we believe that teachers should use a variety of testing instruments, including performance assessments, to get a rich picture of what students know and can do in science.

Now that you have had hands-on experience with a performance assessment, we will, in the next chapter, construct a definition of performance assessment by drawing on what you have experienced.

Summary

In this chapter, you explored the differences between multiple-choice tests and performance assessments. You discovered that differences may exist in the knowledge assessed by the two ways of testing. Multiple-choice tests tend to emphasize recall of facts and concepts; performance assessments invite students to solve a problem, using concrete materials that react to their actions and allow for alternative solution paths. Performance tests are usually engaging and fun to take, less stressful than multiple-choice tests because they are fully contextualized and concrete. Students have told us they consider performance assessments to be more fair in assessing what they have learned in hands-on science, because they look and feel like the learning activities they do in their labs. Furthermore, performance assessments encourage teachers to teach not only facts and concepts but also how to do science.

At the conclusion of this chapter, we considered the differences between the two kinds of test by comparing their component parts—question or task, response format, and scoring system.

Defining Performance Assessment

Overview

In this chapter, we define performance assessment by teasing apart its components and describing the characteristics of each. You will recall from the last chapter that the components are the *task,* the *response* format, and the *scoring* system.

Then we will consider how the structure of the task and the response format control just what can be scored in the assessment. Assessments can be more or less structured, depending on their purpose.

What Is a Performance Assessment?

Think and Respond

Now that you've experienced a performance assessment, how would you define *performance assessment?* Try constructing a definition for yourself.

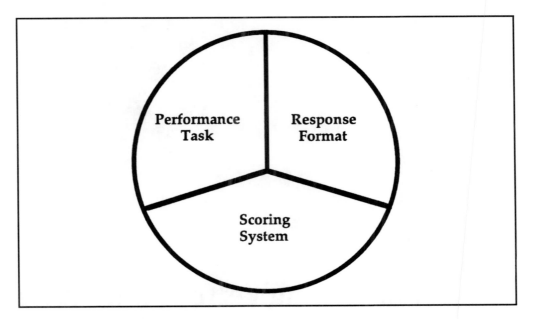

Figure 3.1. Components of a Performance Assessment

_____

_____

_____

One way to define it is simply as the three elements it comprises: a performance task, a response format, and a scoring system, with specific attributes for each component (Figure 3.1). We'll look at each of these components in detail.

Performance Task

In a performance assessment, students are given an invitation to solve a problem or conduct an investigation—this is their task—and they are provided concrete materials to do so. These materials are a key element in performance assessment because they not only provide a context for the problem but also react to the actions of students, thereby giving them feedback that might, in itself, help them to solve the problem.

For example, with the House assessment, the task was to signal a friend using a bright and dim signal light. You tried different configurations of circuits to make your signals, letting the intensity of the lightbulb confirm your hunch about how to solve the problem. You might have used two batteries and one bulb to make your bright signal and one battery and one bulb for dim; or you might have used one battery with one bulb for bright and one battery with two bulbs for dim. Either solution turns out to be successful. This is an important feature of a performance assessment—and it mirrors scientific inquiry itself.

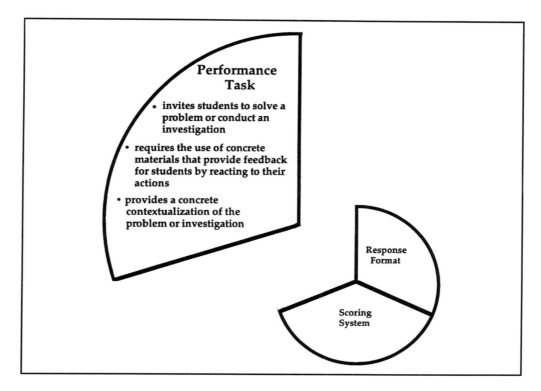

Figure 3.2. Characteristics of a Performance Task

There may be many different solution paths equally appropriate in response to any one task, and you are challenged to find one that is reasonable and effective. Remember that a performance assessment asks students to *perform an investigation* and then evaluates that performance in terms of both procedures and outcomes. Figure 3.2 summarizes the characteristics of the performance task.

Response Format

The second component of a performance assessment, the response format, asks students to communicate their findings in a particular way. They might be asked to decide for themselves how to summarize findings or they might be directed to use a table, a graph, or pictures. For example, in the House assessment you were asked to draw the circuits you made for the dimmer *visit* signal and the brighter *call* signal. On the other hand, students might be asked to list the procedures they used or respond to short-answer probes. (You'll have a chance to experience an assessment with this response format in Chapter 4.) Some performance assessments ask students to explain to a friend how they approached a problem and what they found out. Frequently, students are asked to justify their answers or display the evidence used to reach their conclusions. A summary of the characteristics of a performance assessment's response format is shown in Figure 3.3.

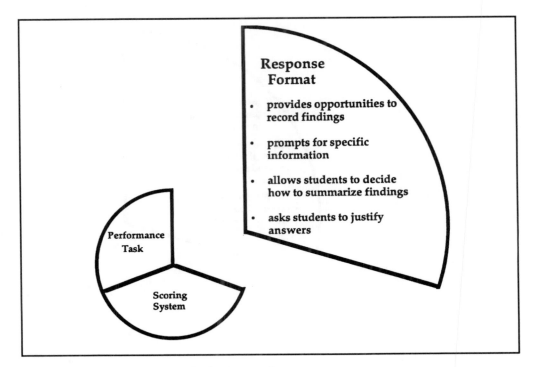

Figure 3.3. Characteristics of a Response Format

Scoring System

Remember, without an established scoring system you simply don't have a performance assessment because you don't know what counts as appropriate procedures and acceptable conclusions or solutions. Indeed, scoring really defines what you are assessing. Without a scoring system, you might have a wonderful hands-on learning activity, but if you don't have a way to evaluate the performance, you can't consider it an assessment.

Throughout this book, we will be showing you different kinds of scoring systems that are appropriate with different genres of investigations. For now, however, we can identify some common characteristics all scoring systems must have.

First of all, they need to take into consideration the variety of solution paths students might use. For each assessment, there could be many equally good ways to solve the problem or conduct an investigation, and the scoring system needs to allow for this creativity on the students' part. Furthermore, some solution paths may be less accurate, less efficient, or less reasonable, and the scoring system needs to judge those solutions appropriately.

Scoring needs to provide the teacher with useful information about what students know and can do, for both accountability and instructional purposes. The performance assessment should help teachers know what concepts are not understood or where misconceptions lie in the minds of their students.

The score form should be so well focused that two teachers, scoring independently, can come up with virtually the same scores. This is called

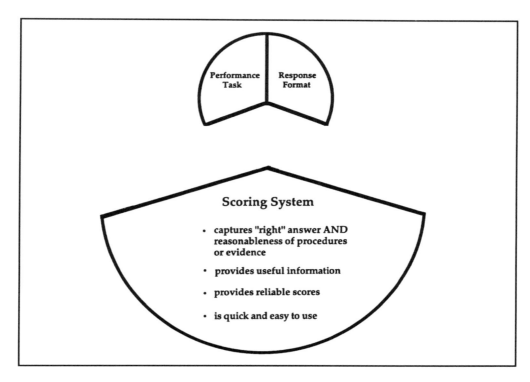

Figure 3.4. Characteristics of a Scoring System

interrater reliability and is an important quality of a performance assessment, which we will discuss further in Chapter 7.

Finally, the score form needs to be quick and easy to use (with some training, of course). After scoring about 10 students, you should be up to speed—and scoring should take no longer than 1 to 2 minutes per student.

Figure 3.4 gives a summary of the characteristics of a performance assessment's scoring system.

Now that we have described a performance assessment as comprising a performance task, a response format, and a scoring system, can we expect all performance assessments with these elements to give us the same quality of information about students' problem-solving ability? Unfortunately not. There are other dimensions to performance assessments that need to be addressed before going any further. One particularly important dimension involves just how the assessment itself is structured. In other words, how much thinking and problem solving are being called for on the part of the student?

Degrees of Assessment Structure

The degree to which the task and the response format facilitate or constrain the alternative procedures that students may use to investigate and solve a problem is called *assessment structure*. Assessment structure answers the question, "How much thinking and problem solving is the student's performance going to demonstrate as he or she addresses the problem?"

If the assessment is highly structured (see below), it might give the student step-by-step instructions to follow (like a recipe in a cookbook). The response format might provide a data table that simply needs to be filled in with student observations. In this case, scoring would be limited to an evaluation of how well the student followed directions and then drew conclusions. There is an axiom in testing that says, "What you test is what you get." You can't evaluate a student's problem-solving procedures if you spoon-feed the procedures to him or her.

Assessment Structure

Degree of Structure	Student Finds Problem	Student Finds Procedures	Student Finds Outcomes
High (Recipe)	No	No	Yes
Low (Discovery)	No (sometimes Yes)	Yes	Yes

On the other hand, the task and response format may be more unstructured, giving the student the opportunity to design and carry out an investigation and then organize and summarize findings—presenting what was *discovered*. Obviously, the more unstructured the assessment, the more information you will gain through your scoring system about what the student knows and can do in science.

The structure of an assessment might depend on factors such as the age of the students and their situation. For example, if students have an extended time to work, they might be asked to identify the problem, design and carry out procedures, and then reach conclusions. With younger children or limited time, however, more structure can be provided.

This chapter has focused on defining the elements that make up a performance assessment and exploring how the structure of each of these elements determines the information you can get from them. In the next chapter, you will have hands-on experience with a low structured assessment called Paper Towels and its specially designed *procedure-based* scoring system.

Summary

A performance assessment can be defined as (a) a task that invites students to solve a concrete, well-contextualized problem, using actual materials that react to the actions of the students; (b) a response format that provides students the opportunity to record findings (in tables, graphs, pictures, or words); and

(c) a scoring system that captures not just the correct answer but also the reasonableness of the procedures used to arrive at an answer.

Performance assessments can be more or less structured, depending on the purpose for which the assessment is given. An assessment with high structure provides students with not only the problem but also step-by-step procedures for solving the problem. Students then draw conclusions based on their findings. This structure limits the information that can be inferred about the students' problem-solving ability. An assessment with low structure, however, gives students more opportunity to think and create. They may be asked to design and carry out an investigation, organize and summarize findings, and then draw conclusions. Obviously, this type of assessment offers more information about what students know and can do in science.

How Do You Score
Student Performance?

Overview

In this chapter, you will experience another performance assessment, one that asks students to design and carry out an investigation to determine which of three brands of paper towels absorbs the most and least water. This assessment, called Paper Towels, represents a category of assessment that we call a *comparative investigation*, because we want students to compare the towels on one attribute: absorbency. This assessment has been used with students from grades 3 through 8, primarily assessing procedural ability.

You will conduct the investigation yourself, reporting your procedures and findings in an *assessment notebook*. Then you will be introduced to the scoring system that was designed for this assessment. We use a *procedure-based* score form because we want to evaluate the procedures students used to conduct their investigations as well as the conclusions reached in the investigations.

Gather the following materials for the Paper Towels assessment:

PAPER TOWELS MATERIALS LIST

- Kitchen timer

- Eyedropper

- Scissors

- Measuring cup (1 pint or 500 ml), filled with water

- Magnifying glass or magnifying lens

- 3 graduated measuring glasses, approximately 7 oz
  (You can write a scale on each of 3 plastic cups if you want.)

- Food (diet) scale, postage scale, or spring scale

- 12-inch ruler

- 3 rolls of paper towels, each a different brand with distinguishing markings (e.g., solid white, blue designs, etc.). One brand should be a generic, 1-ply "cheap" white paper towel; the others should be more expensive, heavy, brand-name paper towels. There may be very little difference in the absorbency of some brands of paper towels, so it is important to include at least 2 rolls with major differences. (For our purposes here, you need only a few sheets of each.)

**Optional Equipment
(included in the full assessment)**

- Tweezers

- 3 petri dishes with lids

- 3 plastic or foil trays (12 in × 7 in, about ¾-in deep)

- 2 funnels (1 small, 1 medium sized)

- 2 beakers (200 ml and 50 ml)

In the last chapter we described assessments with both high and low structure. Recall that the amount of information you can get about what your students can do in science depends on how much investigating and problem solving you allow them to demonstrate. The next assessment you will do is a classic task familiar to schools in both the United States and England. The version of this task we have used in our research, called Paper Towels, is an example of an assessment with low structure, inviting students to design and carry out an investigation to determine which of three brands of paper towels absorbs the most and least water. The students' procedures and findings are then scored with a procedure-based scoring system.

Experiencing the Paper Towels Assessment

Collect the materials listed on page 27. Notice that some of the equipment is listed as optional. We provide students with a variety of relevant lab equipment so they have an opportunity to choose what they will need instead of being led into a particular investigation by the equipment. For the same reason, we also provide some irrelevant equipment. Because, realistically, you might not want to assemble all the equipment, some relevant and some not, we've listed "optional" equipment.

Array the equipment as shown in Figure 4.1 in a semicircle in front of you, the "student." All equipment should be within easy reach, and no one piece of equipment should be in a position of more importance than the others. In other words, we do not want to lead the student into conducting the investigation in a certain manner.

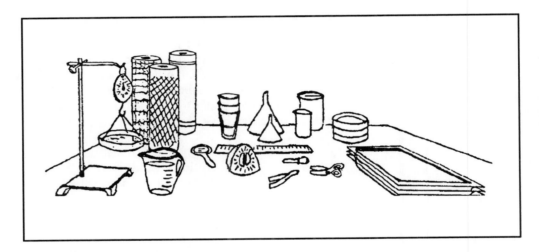

Figure 4.1. Paper Towels Setup

When you have assembled all your materials, follow the directions on pages 29-33, doing the assessment yourself. (We used white, blue, and yellow paper towels, so those are the colors mentioned in the assessment pages. If you use different colors, you will of course need to change the colors in the instructions to match your new colors.) When you finish doing the assessment yourself, keep the equipment handy. We want you to also observe someone else doing the assessment—but we'll tell you about that when the time comes.

Go ahead and do the assessment now.

Components of Paper Towels

Now that you have done the assessment, let's think about Paper Towels in terms of what we have already learned. The assessment is composed of a *task* that asked you to find out, "Which brand of paper towels absorbs the most and

Paper Towels

Name _____

You have three different kinds of paper towels in front of you and some equipment for doing scientific experiments.

Problems:

1.   Find out which paper towel can hold, soak up or absorb the <u>most</u> water.

2.   Find out which paper towel can hold, soak up or absorb the <u>least</u> water.

Look at each piece of equipment. Think about how you might use some of it to do an experiment to solve the problems. You don't need to use all the equipment.

When you are finished you will be asked to write what you did so one of your friends can repeat the experiment exactly as you did it. You may want to keep notes on a sheet of paper to help you remember what you did and what you found out.

Figure 4.2. Paper Towels Performance Assessment (1 of 5)
Brown and Shavelson, *Assessing Hands-On Science: A Teacher's Guide to Performance Assessment.* Copyright 1996, Corwin Press, Inc.

NOTES

RESULTS: When you think you know which paper towel can hold, soak up or absorb the most water and least water, write **"most"** and **"least"** beside the name of the towel.

White _____   Blue _____   Yellow _____

How did you know from the experiment which paper towel holds, soaks up or absorbs the **most** water and which paper towel holds, soaks up or absorbs the **least** water?

Most

Least

Figure 4.2. Paper Towels Performance Assessment (2 of 5)
Brown and Shavelson, *Assessing Hands-On Science: A Teacher's Guide to Performance Assessment.* Copyright 1996, Corwin Press, Inc.

Paper Towels Notebook
(page 1)

Name _____

       Scientists keep notebooks when they do experiments to remind them of what they did. Also, it tells other scientists the steps in the experiment so they can use the notebook to repeat the experiment.

       In the spaces provided on the following pages, write your scientific notebook. Describe in detail what you did at each step in your experiment so one of your classmates can do the experiment exactly as you did it.

Please turn to page 2

Figure 4.2. Paper Towels Performance Assessment (3 of 5)

Paper Towels Notebook
(page 2)

A.   Steps in Experiment: Please number each step in order.

Step   **What you did**

<u>1.</u> _____

_____

_____

_____

_____

_____

_____

_____

_____

_____

_____

_____

_____

_____

_____

Please turn to page 3

Figure 4.2. Paper Towels Performance Assessment (4 of 5)
Brown and Shavelson, *Assessing Hands-On Science: A Teacher's Guide to Performance Assessment.* Copyright 1996, Corwin Press, Inc.

Paper Towels Notebook
(page 3)

B. **Here are some questions about your experiment. Answer each of the questions "yes" or "no".**

1. Were all the paper towels the same size?   _____

2. Were all the paper towels completely wet?   _____

3. Did you use the same amount of water to get each
    paper towel wet?   _____

4. Did you let each towel soak in the water for the
    same amount of time?   _____

C. **How did you know from the experiment which paper towel holds, soaks up or absorbs the <u>most</u> water and which paper towel holds, soaks up or absorbs the <u>least</u> water?**

Most _____

_____

Least _____

_____

D. **Francisco thinks all of the paper towels <u>must be completely wet</u> before you can decide which paper towel holds the most water and which holds the least. Sally <u>does not</u> think the paper towels have to be <u>completely wet</u>. What do you think?**

_____

_____

_____

Figure 4.2. Paper Towels Performance Assessment (5 of 5)

Brown and Shavelson, *Assessing Hands-On Science: A Teacher's Guide to Performance Assessment.* Copyright 1996, Corwin Press, Inc.

least water?" The materials provided reacted to the actions you took to conduct the investigation. The *response format* asked you to list the steps you took in your investigation and what you found out; then you were given some additional questions to answer to further clarify what you did. Although the problem was set for you—which of three paper towels absorbs the most and least water—the task was unstructured. You were given opportunity to design and carry out an investigation for yourself. Then you drew a conclusion based on your own investigation, deciding which paper towels held the most and least water. The response format was also unstructured—you were not asked to fill in parts of step-by-step instructions—rather you had to decide what to tell others about your procedure and write that down. After describing your procedure, you were directly asked some additional questions (a more structured portion of the response format). For example, on page 3 of the Notebook, you were asked, "Were all the paper towels completely wet?" This was done because often students don't clearly state in their procedures whether the towels were saturated with water.

Scoring the Paper Towels Assessment

First, consider for yourself how you would score performance on this assessment. What makes for a *scientific* investigation? What should count as reasonable scientific procedures in the investigation?

To help you think about scoring, we encourage you to find a student—an elementary-school student, a teenager, or an adult—and watch that person do the investigation. Because this assessment measures general scientific thinking skill rather than some knowledge in a content domain (e.g., electricity), it is not necessary for your student to have completed a particular instructional unit. Observe the student and evaluate how well he or she performs the task. Then come up with a score and justification for the score. Test your student now, using pages 35 to 39. Record your observations on page 41.

How Did You Score This Assessment?

What score did you give your student? What were your justifications for that score? When we have had teachers at our workshops observe students doing Paper Towels, scoring them as you did, the teachers were always amazed at how differently they judge and score the assessment. For the same student's performance, we see scores such as A, C+, Satisfactory, 8, and ✓. Why is there such variety in the scores? Aside from using different scales, each teacher focuses on different aspects of the performance, with different criteria and different weights for each aspect.

Common sense tells us that for an assessment to be trustworthy (or reliable, as testing experts would say), it needs to be scored consistently no matter who scores the test. Obviously, what we need is both a predetermined set of criteria on which to judge the performance and a score form that translates those criteria into observable chunks. The scorer needs to know what to look for and then how to assign points or a grade to the performance as a whole, based on analyzing particular parts of the performance.

Paper Towels

Name _____

You have three different kinds of paper towels in front of you and some equipment for doing scientific experiments.

Problems:

1.   Find out which paper towel can hold, soak up or absorb the <u>most</u> water.

2.   Find out which paper towel can hold, soak up or absorb the <u>least</u> water.

Look at each piece of equipment. Think about how you might use some of it to do an experiment to solve the problems. You don't need to use all the equipment.

When you are finished you will be asked to write what you did so one of your friends can repeat the experiment exactly as you did it. You may want to keep notes on a sheet of paper to help you remember what you did and what you found out.

Figure 4.3. Paper Towels Performance Assessment (1 of 5)

Brown and Shavelson, *Assessing Hands-On Science: A Teacher's Guide to Performance Assessment.* Copyright 1996, Corwin Press, Inc.

NOTES

RESULTS: When you think you know which paper towel can hold, soak up or absorb the most water and least water, write **"most"** and **"least"** beside the name of the towel.

White _____   Blue _____   Yellow _____

How did you know from the experiment which paper towel holds, soaks up or absorbs the **most** water and which paper towel holds, soaks up or absorbs the **least** water?

Most

Least

Figure 4.3. Paper Towels Performance Assessment (2 of 5)

Brown and Shavelson, *Assessing Hands-On Science: A Teacher's Guide to Performance Assessment.* Copyright 1996, Corwin Press, Inc.

Paper Towels Notebook
(page 1)

Name _____

Scientists keep notebooks when they do experiments to remind them of what they did. Also, it tells other scientists the steps in the experiment so they can use the notebook to repeat the experiment.

In the spaces provided on the following pages, write your scientific notebook. Describe in detail what you did at each step in your experiment so one of your classmates can do the experiment exactly as you did it.

Please turn to page 2

Figure 4.3. Paper Towels Performance Assessment (3 of 5)

Paper Towels Notebook
(page 2)

A.   Steps in Experiment: Please number each step in order.

Step   **What you did**

1. _____

_____

_____

_____

_____

_____

_____

_____

_____

_____

_____

_____

_____

_____

_____

_____

Please turn to page 3

Figure 4.3. Paper Towels Performance Assessment (4 of 5)

Paper Towels Notebook
(page 3)

B.   Here are some questions about your experiment. Answer each of the questions "yes" or "no".

1.   Were all the paper towels the same size?   _____

2.   Were all the paper towels completely wet?   _____

3.   Did you use the same amount of water to get each paper towel wet?   _____

4.   Did you let each towel soak in the water for the same amount of time?   _____

C.   How did you know from the experiment which paper towel holds, soaks up or absorbs the <u>most</u> water and which paper towel holds, soaks up or absorbs the <u>least</u> water?

Most _____

_____

Least _____

_____

D.   Francisco thinks all of the paper towels <u>must be completely wet</u> before you can decide which paper towel holds the most water and which holds the least. Sally <u>does not</u> think the paper towels have to be <u>completely wet</u>. What do you think?

_____

_____

Figure 4.3. Paper Towels Performance Assessment (5 of 5)
Brown and Shavelson, *Assessing Hands-On Science: A Teacher's Guide to Performance Assessment.* Copyright 1996, Corwin Press, Inc.

Paper Towels Investigation
Observer's Scoring Sheet

My score for this "student"_____

Justification: _____

_____

_____

_____

_____

_____

_____

_____

What did you consider the important aspects of the performance?

_____

_____

_____

_____

_____

_____

_____

_____

_____

_____

Figure 4.4. Observer's Scoring Sheet
Brown and Shavelson, *Assessing Hands-On Science: A Teacher's Guide to Performance Assessment.* Copyright 1996, Corwin Press, Inc.

Developing a Scoring System for Paper Towels

To develop this score form, we watched hundreds of students do the investigation and we designed the score form to allow for all the possible combinations of performances we observed. Of course, room always needs to be left for originality—students never fail to surprise you—so we included a space for "other" ways of determining results. Then we set up criteria for grading the performance according to the reasonableness of the various solutions. Basically, we wanted to identify the following components of the student's performance:

Procedure

1. The *method for getting the towel wet*
2. *Saturation*—that is, did the student get the towels completely wet
3. How the student *determined results*
4. Was the student *careful in saturation and measuring*?

Results

5. Did the student get the *correct result*?

A key to this assessment is the student's realization that each towel had to be saturated for a fair comparison to be made between them. In scoring performance then, one crucial question is, Did the student control for the variable of saturation? Another important aspect of performance is whether the student used a method for determining the results that logically followed from the way he or she got the towels wet in the first place. For example, a student could very carefully measure equal amounts of water into three cups, saturate a towel in each cup, remove the towel, and compare how much water remained in the cups. Notice that a logical progression follows between the method of wetting the towels and that of determining the results. In contrast, if the student hadn't originally put equal amounts of water in the cups, comparing water remaining after the towels were removed would not be "fair" and might lead to the wrong conclusion.

The Paper Towels Score Form

On page 43 is the score form that our research team developed for Paper Towels (Baxter, Shavelson, Goldman, & Pine, 1992). You will see that all the aspects of the student's investigation mentioned above are listed so you can circle what the student did. Then you find the row at the bottom that describes the grade given to various investigations.

Scoring Alonzo's Investigation

The best way to see how this score form works is to score some students' investigations. Let's begin by observing Alonzo's investigation. (Photos 1

Paper Towels Score Form

Student: _____Observer/Scorer_____Score_____

1. Method for Getting Towel Wet

    A. Container   **B.** Drops   **C.** Tray (surface)   **D.** No Method
        Put water in/put towel in   Towel on tray/pour water on
        Put towel in/pour water in   Pour water on tray/put towel in
            1 pitcher or 3 beakers/glasses

2. Saturation   **A.** Yes   **B.** No   **C.** Controlled
                                                  (same amount of water-all towels)

3. Determine Result

        A. Weigh towel

        B. Squeeze towel/measure water (weight or volume)

        C. Measure water in/out

        D. Count # drops until saturated

        E. Irrelevant measurement

        (i.e. time to soak up water, see how far drops spread out, feel thickness)

        F. Other _____

4. Care in saturation &/or measuring   Yes   No   A little sloppy   (+/-)

5. Correct result   Most   Least

Grade	Method	Saturate	Determine Result	Care in Measuring	Correct Answer
A	Yes	Yes	Yes	Yes	Both
B	Yes	Yes	Yes	No	One or Both
C	Yes	Controlled	Yes	Yes/No	One or Both
D	Yes	No *or*	Inconsistent	Yes/No	One or Both
F	Inconsistent *or*	No	Irrelevant	Yes/No	One or Both

Figure 4.5. Paper Towels Score Form

Photo 1: For his investigation, Alonzo first tears off a paper towel and dunks it in the container of water.

through 3 show key aspects of his performance.) First, he tore a paper towel off one roll. Then, he folded it and dunked it in the container of water, making sure it was completely saturated. Next, he carefully squeezed the towel, collecting the water in a cup. He repeated this for all three towels and then compared the cups to see which had the most and the least water.

Figure 4.6 on page 46 shows how Alonzo's assessment would be scored if we observed him doing it directly. We have circled *1A*, put towel in container; *2A*, yes, he did saturate the towels; *3B*, he squeezed the towels into cups and then compared the volume of water in each cup to determine which held the least and the most; *4A*, yes, he was reasonably careful; and *5*, he got the right answers, most and least. Looking at the bottom of the score form, we can match up what he did and give him a score. Looking across the row marked *A: Yes*, he used a consistent (same for all three towels) and reasonable way of getting the towels wet; yes, he saturated the towels; yes, he used a logical method of determining results; yes, he used normal care in saturating and measuring; and yes, he got the correct result. He earns an A.

Practice Scoring

Are you beginning to see how this score form works? For the next two students' investigations, go ahead and score each student on your own before you read through our description of the scoring; this will give you some practice. First, meet Charlie and score his performance on page 47 (see Photos 4 through 6).

Photo 2: Then he squeezes the water held in the paper towel in a graduated plastic cup.

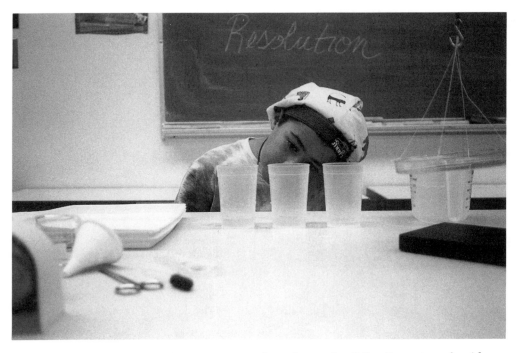

Photo 3: After following the same procedure for each of the three towels, Alonzo compares the amount of water squeezed into each cup, correctly noting which held the most and least.

Paper Towels Score Form

Student: ___*Alonzo*_____ Observer/Scorer_____ Score _A___

1. Method for Getting Towel Wet

A. Container (circled)
Put water in/put towel in
Put towel in/pour water in
    1 pitcher or 3 beakers/glasses

B. Drops   **C.** Tray (surface)   **D.** No Method
                    Towel on tray/pour water on
                    Pour water on tray/put towel in

2. Saturation   **A. Yes** (circled)   **B.** No   **C.** Controlled
                                                          (same amount of water-all towels)

3. Determine Result

  A. Weigh towel

  B. Squeeze towel/measure water (weight or volume) (circled)

  C. Measure water in/out

  D. Count # drops until saturated

  E. Irrelevant measurement

       (i.e. time to soak up water, see how far drops spread out, feel thickness)

  F. Other _____

4. Care in saturation &/or measuring   Yes (circled)   No   A little sloppy   (+/-)

5. Correct result   Most (circled)   Least (circled)

Grade	Method	Saturate	Determine Result	Care in Measuring	Correct Answer
A (circled)	Yes ✓	Yes ✓	Yes ✓	Yes ✓	Both ✓
B	Yes	Yes	Yes	No	One or Both
C	Yes	Controlled	Yes	Yes/No	One or Both
D	Yes	No *or*	Inconsistent	Yes/No	One or Both
F	Inconsistent *or*	No	Irrelevant	Yes/No	One or Both

Figure 4.6. Paper Towels Score Form
Brown and Shavelson, *Assessing Hands-On Science: A Teacher's Guide to Performance Assessment.* Copyright 1996, Corwin Press, Inc.

Paper Towels Score Form

Student: _____*Charlie*_____Observer/Scorer_____Score_____

1. **Method for Getting Towel Wet**

    A. Container
       Put water in/put towel in
       Put towel in/pour water in
          1 pitcher or 3 beakers/glasses
    B. Drops
    C. Tray (surface)
       Towel on tray/pour water on
       Pour water on tray/put towel in
    D. No Method

2. **Saturation**   A. Yes   B. No   C. Controlled
                                          (same amount of water-all towels)

3. **Determine Result**

    A. Weigh towel

    B. Squeeze towel/measure water (weight or volume)

    C. Measure water in/out

    D. Count # drops until saturated

    E. Irrelevant measurement

       (i.e. time to soak up water, see how far drops spread out, feel thickness)

    F. Other _____

4. **Care in saturation &/or measuring**   Yes   No   A little sloppy   (+/-)

5. **Correct result**   <u>Most</u>   Least

Grade	Method	Saturate	Determine Result	Care in Measuring	Correct Answer
A	Yes	Yes	Yes	Yes	Both
B	Yes	Yes	Yes	No	One or Both
C	Yes	Controlled	Yes	Yes/No	One or Both
D	Yes	No *or*	Inconsistent	Yes/No	One or Both
F	Inconsistent *or*	No	Irrelevant	Yes/No	One or Both

Figure 4.7. Paper Towels Score Form
Brown and Shavelson, *Assessing Hands-On Science: A Teacher's Guide to Performance Assessment.* Copyright 1996, Corwin Press, Inc.

Photo 4: Charlie places a paper towel on each of the three trays and then, using the eyedropper, begins to count the drops he places on the first towel.

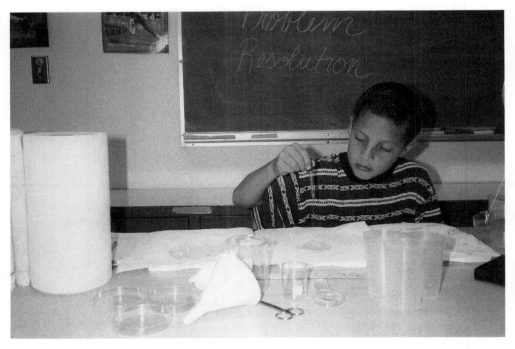

Photo 5: He soon decides that counting drops will take forever! So on the next two towels he squirts a few droppers full on each one.

Photo 6: Charlie determines which towel holds the most and least water by looking under each one to see how much water leaked through to the tray. He correctly identifies the "most" and "least" absorbency.

Scoring Charlie's Investigation

Charlie used the eyedropper as his method of getting the towels wet. He started counting the drops he placed on the first towel but soon realized that method was too laborious, so he gave the other two towels a few squirts from the eyedropper. He obviously did not saturate the towels. His method of determining which held the most and the least water was to hold up the towels to see how much water was being held (the old "eyeball" method). Obviously, this investigation does not offer a viable solution to the problem.

Now look at the bottom of the page to assign Charlie a grade. He used the eyedropper to wet the towels, but he did not, in fact, saturate the towels; then he used an irrelevant method to determine his results (he looked under each towel to see how much water had soaked through). His investigation is scored with an F. Our score form for Charlie is on page 50.

Scoring Amy's Investigation

Now score Amy's investigation for yourself (see Photos 7 through 9) using the score form on page 51 and then check your score with ours found on page 54.

Amy placed a towel in a tray and poured 30 ml of water in the center of the towel (it was not enough to saturate the towel) and weighed the towel on the scale. She repeated this procedure for each of the towels and then compared weights. We circle *1C*, towel on tray/pour water on tray. For *2* we circle C

Paper Towels Score Form

Student: ____*Charlie*_____Observer/Scorer_____Score__F__

1. **Method for Getting Towel Wet**

   A. Container   B. Drops   C. Tray (surface)   D. No Method
       Put water in/put towel in   *Inconsistent*   Towel on tray/pour water on
       Put towel in/pour water in   Pour water on tray/put towel in
        1 pitcher or 3 beakers/glasses

2. **Saturation**   A. Yes   B. No   C. Controlled
                                             (same amount of water-all towels)

3. **Determine Result**

   A. Weigh towel

   B. Squeeze towel/measure water (weight or volume)

   C. Measure water in/out

   D. Count # drops until saturated

   E. Irrelevant measurement

        (i.e. time to soak up water, see how far drops spread out, feel thickness)

   F. Other _____

4. **Care in saturation &/or measuring**   Yes   No   A little sloppy   (+/-)

5. **Correct result**   Most   Least

Grade	Method	Saturate	Determine Result	Care in Measuring	Correct Answer
A	Yes	Yes	Yes	Yes	Both
B	Yes	Yes	Yes	No	One or Both
C	Yes	Controlled	Yes	Yes/No	One or Both
D	Yes	No *or*	Inconsistent	Yes/No	One or Both
F	Inconsistent *or*	No	Irrelevant	Yes/No	One or Both

Figure 4.8. Paper Towels Score Form
Brown and Shavelson, *Assessing Hands-On Science: A Teacher's Guide to Performance Assessment.* Copyright 1996, Corwin Press, Inc.

Paper Towels Score Form

Student: __*Amy*_____ Observer/Scorer_____ Score_____

1. Method for Getting Towel Wet

    A. Container   **B.** Drops   **C.** Tray (surface)   **D.** No Method
        Put water in/put towel in   Towel on tray/pour water on
        Put towel in/pour water in   Pour water on tray/put towel in
             1 pitcher or 3 beakers/glasses

2. Saturation   **A.** Yes   **B.** No   **C.** Controlled
                                                          (same amount of water-all towels)

3. Determine Result

         A.   Weigh towel

         B.   Squeeze towel/measure water (weight or volume)

         C.   Measure water in/out

         D.   Count # drops until saturated

         E.   Irrelevant measurement

              (i.e. time to soak up water, see how far drops spread out, feel thickness)

         F.   Other _____

4. Care in saturation &/or measuring   Yes   No   A little sloppy   (+/-)

5. Correct result   Most   Least

Grade	Method	Saturate	Determine Result	Care in Measuring	Correct Answer
A	Yes	Yes	Yes	Yes	Both
B	Yes	Yes	Yes	No	One or Both
C	Yes	Controlled	Yes	Yes/No	One or Both
D	Yes	No *or*	Inconsistent	Yes/No	One or Both
F	Inconsistent *or*	No	Irrelevant	Yes/No	One or Both

Figure 4.9. Paper Towels Score Form

Brown and Shavelson, *Assessing Hands-On Science: A Teacher's Guide to Performance Assessment.* Copyright 1996, Corwin Press, Inc.

Photo 7: After placing a towel in the center of each tray, Amy carefully measures 30 ml of water into a glass beaker.

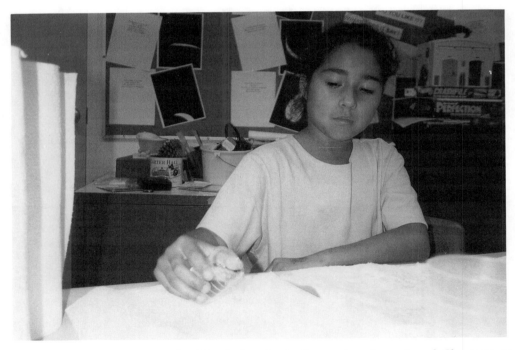

Photo 8: Then Amy pours the 30 ml of water in the center of one towel. She repeats this process for the other two towels. (Thirty ml of water is not enough to fully saturate the towel.)

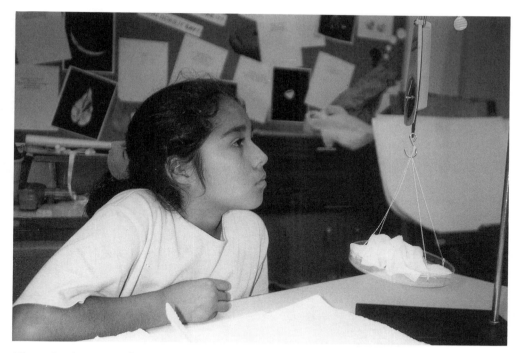

Photo 9: Amy weighs each towel, records the weights, and correctly identifies "most" and "least" absorbency.

(controlled), because although she did not saturate the towels, she did attempt to control some aspect of the application of water. Then we circle 3A (weigh towel) as her method of determining the results. She used normal care (circle 4, yes) and got the correct results (circle 5, most and least). When we look at the bottom of the page to see what score to give her, we find that she cannot earn an A or B because she did not saturate the towels. She does earn a C, however, because she controlled the amount of water put on each towel and used a reasonable method to determine her results.

Scoring in the Real World

Are you becoming more comfortable with the score form? We have found that by the time scorers practice with about 10 students, they are able to score easily, quickly, and reliably.

We observed Alonzo, Charlie, and Amy doing investigations and scored their performance directly. But how can we possibly do that in a classroom with dozens of children? Fortunately, research has shown that a student's *assessment notebook*, like the one in which you recorded your procedures on pages 31-33, can take the place of direct observation by the teacher (Baxter & Shavelson, 1994; Baxter et al., 1992). In other words, you can trust the students' assessment notebooks to give you scores that are comparable to what they would earn if you were directly observing them. The only difference is that you really cannot tell whether the student used *normal care* in conducting the

Paper Towels Score Form

Student: ___*Amy*_____Observer/Scorer_____Score__*C*___

1. Method for Getting Towel Wet

A. Container   B. Drops   (C. Tray (surface))   D. No Method
     Put water in/put towel in   Towel on tray/pour water on
     Put towel in/pour water in   Pour water on tray/put towel in
         1 pitcher or 3 beakers/glasses

2. Saturation   A. Yes   B. No   (C. Controlled)
                                             (same amount of water-all towels)

3. Determine Result

     (A.   Weigh towel)

     B.   Squeeze towel/measure water (weight or volume)

     C.   Measure water in/out

     D.   Count # drops until saturated

     E.   Irrelevant measurement

         (i.e. time to soak up water, see how far drops spread out, feel thickness)

     F.   Other _____

4. Care in saturation &/or measuring   (Yes)   No   A little sloppy   (+/-)

5. Correct result   (Most)   (Least)

Grade	Method	Saturate	Determine Result	Care in Measuring	Correct Answer
A	Yes	Yes	Yes	Yes	Both
B	Yes	Yes	Yes	No	One or Both
(C)	Yes ✓	Controlled ✓	Yes ✓	Yes/No ✓	One or Both ✓
D	Yes	No *or*	Inconsistent	Yes/No	One or Both
F	Inconsistent *or*	No	Irrelevant	Yes/No	One or Both

Figure 4.10. Paper Towels Score Form

Brown and Shavelson, *Assessing Hands-On Science: A Teacher's Guide to Performance Assessment.* Copyright 1996, Corwin Press, Inc.

experiment; so when we score student notebooks, we cross out *4, care in saturation and/or measuring*. Otherwise, this score form is used in the same way.

Scoring an Assessment Notebook

Now we will look at Cecilia's investigation by evaluating her assessment notebook and then we'll score it. Remember that well-designed assessment notebooks allow teachers to score students' investigations after the actual testing period almost as if they were looking over the students' shoulders. Read Cecilia's assessment notebook, on pages 57-59, and see if you can score it yourself, using the score form on page 56. Then look at how your scoring matches up with ours on page 60.

How We Scored Cecilia's Notebook

We read that Cecilia soaked all three towels in water (circle *1A*), saturated them (circle *2A*), and then weighed each one to determine her results (circle *3A*). Remember, we skip *4* because we cannot tell how carefully she carried out some aspects of her investigation. (Did she spill water? Did she leave water in the scale when weighing the next towel?) She got *5* correct (circle *most* and *least*).

Looking at the bottom of the scoring page to grade her assessment, we find that she fits the A row. Furthermore, because she replicated her investigation, we add a plus. Obviously, this fifth grader conducted an excellent investigation.

With practice, the Paper Towels score form allows the teacher to score a student's complex investigation in a consistent way. The criteria used for judging the performance are clearly determined and easy to apply, so scoring is fast and reliable. But what is the benefit to the classroom teacher? In the next chapter, we explore the usefulness of performance assessments in the classroom, using as our example an assessment designed for a commonly taught upper elementary unit called Mystery Powders.

Summary

This chapter has presented hands-on experience with a comparative investigation and its procedure-based scoring system. First, you experienced doing the Paper Towels assessment yourself and then you observed someone else doing the assessment. Next, you learned to score Paper Towels with a scoring system that allowed you to reliably and quickly evaluate a student's investigation, either through direct observation or by examining the student's assessment notebook.

You were reminded that a good scoring system captures not only the answer but also the reasonableness of the procedures a student uses, evaluating many different solution paths. The system should be quick and easy to use (with some training, of course) and should provide the teacher with useful information about what students know and can do in science. It should also clearly define the criteria on which evaluation is based so that different scorers can be consistent in their scoring.

Paper Towels Score Form

Student: __*Cecilia*_____Observer/Scorer_____Score_____

1. Method for Getting Towel Wet

    A. Container   **B.** Drops   **C.** Tray (surface)   **D.** No Method
         Put water in/put towel in   Towel on tray/pour water on
         Put towel in/pour water in   Pour water on tray/put towel in
              1 pitcher or 3 beakers/glasses

2. Saturation   **A.** Yes   **B.** No   **C.** Controlled
                                                 (same amount of water-all towels)

3. Determine Result

        A.   Weigh towel

        B.   Squeeze towel/measure water (weight or volume)

        C.   Measure water in/out

        D.   Count # drops until saturated

        E.   Irrelevant measurement

               (i.e. time to soak up water, see how far drops spread out, feel thickness)

        F.   Other _____

4. Care in saturation &/or measuring   Yes   No   A little sloppy   (+/-)

5. Correct result   Most   Least

Grade	Method	Saturate	Determine Result	Care in Measuring	Correct Answer
A	Yes	Yes	Yes	Yes	Both
B	Yes	Yes	Yes	No	One or Both
C	Yes	Controlled	Yes	Yes/No	One or Both
D	Yes	No *or*	Inconsistent	Yes/No	One or Both
F	Inconsistent *or*	No	Irrelevant	Yes/No	One or Both

Figure 4.11. Paper Towels Score Form
Brown and Shavelson, *Assessing Hands-On Science: A Teacher's Guide to Performance Assessment.* Copyright 1996, Corwin Press, Inc.

NOTES

paper towel

A all white : 27½ grams, 45½ grams : avg 36g

B orange : 55½ g , 41 g : avg : 48.21g.

C blue : 51 g., 48 g. : ave, 49.1g.

$$
\begin{array}{c}
27\;½ \\
45\;½ \\
\hline
73
\end{array}
\qquad
\begin{array}{c}
36.1 \\
2\overline{)73} \\
-6 \\
\hline
13 \\
12
\end{array}
\qquad
\begin{array}{c}
55\;½ \\
41 \\
\hline
96\;½
\end{array}
\qquad
\begin{array}{c}
48.21 \\
2\overline{)96.5} \\
-8 \\
\hline
16 \\
-16
\end{array}
\qquad
\begin{array}{c}
51 \\
48 \\
\hline
99 \\
\;\;49.1 \\
2\overline{)99} \\
-8 \\
\hline
19
\end{array}
$$

RESULTS: When you think you know which paper towel can hold, soak up or absorb the most water and least water, write **"most"** and **"least"** beside the name of the towel.

White _least_   Blue _most_   Yellow _middle_

How did you know from the experiment which paper towel holds, soaks up or absorbs the **most** water and which paper towel holds, soaks up or absorbs the **least** water?

Most fist I Soaked it in water, pulled it out and let all the water not being held drip out twice, then I averaged

Least the same thing as befor

Figure 4.12. Cecilia's Notebook (1 of 3)

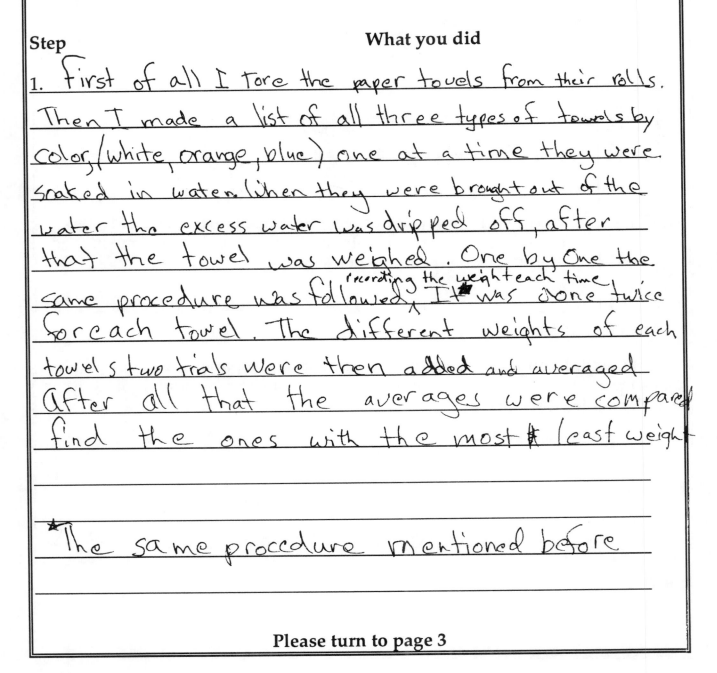

Paper Towels Notebook
(page 2)

A.   Steps in Experiment: Please number each step in order.

Step   **What you did**

1. First of all I tore the paper towels from their rolls.
Then I made a list of all three types of towels by
color, (white, orange, blue) one at a time they were.
soaked in water. When they were brought out of the
water the excess water was dripped off, after
that the towel was weighed. One by One the
                                        recording the weight each time,
same procedure was followed. It was done twice
for each towel. The different weights of each
towels two trials were then added and averaged
after all that the averages were compared
find the ones with the most & least weight

★ The same procedure mentioned before

Please turn to page 3

Figure 4.12. Cecilia's Notebook (2 of 3)
Brown and Shavelson, *Assessing Hands-On Science: A Teacher's Guide to Performance Assessment.* Copyright 1996, Corwin
Press, Inc.

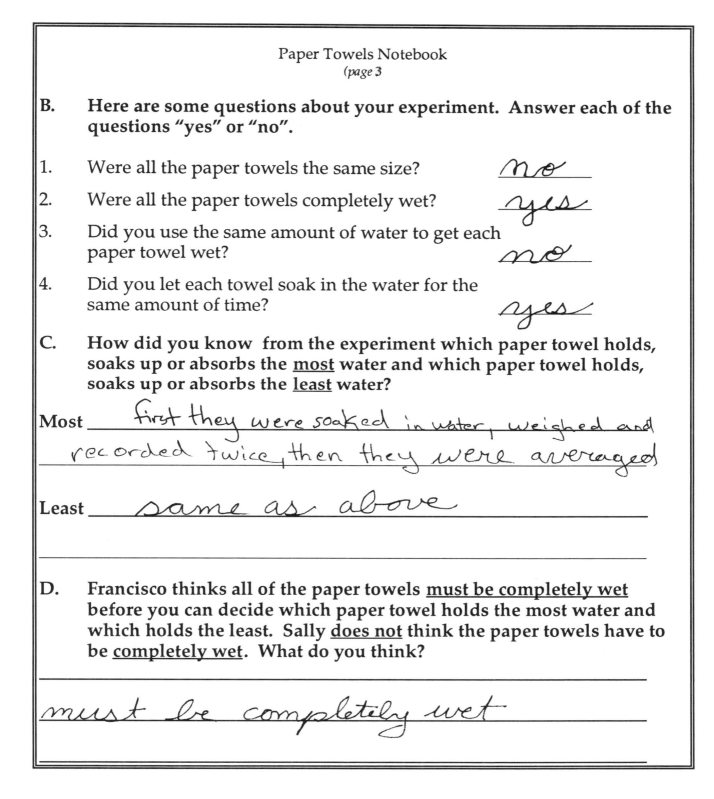

Paper Towels Notebook
(page 3

B.   Here are some questions about your experiment. Answer each of the questions "yes" or "no".

1.   Were all the paper towels the same size?   _no_

2.   Were all the paper towels completely wet?   _yes_

3.   Did you use the same amount of water to get each paper towel wet?   _no_

4.   Did you let each towel soak in the water for the same amount of time?   _yes_

C.   How did you know from the experiment which paper towel holds, soaks up or absorbs the <u>most</u> water and which paper towel holds, soaks up or absorbs the <u>least</u> water?

Most _first they were soaked in water, weighed and recorded twice, then they were averaged_

Least _same as above_

D.   Francisco thinks all of the paper towels <u>must be completely wet</u> before you can decide which paper towel holds the most water and which holds the least. Sally <u>does not</u> think the paper towels have to be <u>completely wet</u>. What do you think?

must be completely wet

Figure 4.12. Cecilia's Notebook (3 of 3)

Brown and Shavelson, *Assessing Hands-On Science: A Teacher's Guide to Performance Assessment.* Copyright 1996, Corwin Press, Inc.

Paper Towels Score Form

Student: __*Cecilia*_____ Observer/Scorer _JB_____ Score _A+_

(replication)

1. **Method for Getting Towel Wet**

    (A.) Container   B. Drops   C. Tray (surface)   D. No Method
          Put water in/put towel in   Towel on tray/pour water on
          Put towel in/pour water in   Pour water on tray/put towel in
            1 pitcher or 3 beakers/glasses

2. **Saturation**   (A. Yes)   B. No   C. Controlled
                                                            (same amount of water-all towels)

3. **Determine Result**

    (A.   Weigh towel)

    B.   Squeeze towel/measure water (weight or volume)

    C.   Measure water in/out

    D.   Count # drops until saturated

    E.   Irrelevant measurement

              (i.e. time to soak up water, see how far drops spread out, feel thickness)

    F.   Other _____

4. ~~Care in saturation &/or measuring~~   ~~Yes   No   A little sloppy~~   (+/-)

5. **Correct result**   (Most)   (Least)

Grade	Method	Saturate	Determine Result	Care in Measuring	Correct Answer
(A)	Yes ✓	Yes ✓	Yes ✓	Yes ✓	Both ✓
B	Yes	Yes	Yes	No	One or Both
C	Yes	Controlled	Yes	Yes/No	One or Both
D	Yes	No *or*	Inconsistent	Yes/No	One or Both
F	Inconsistent *or*	No	Irrelevant	Yes/No	One or Both

Figure 4.13. Paper Towels Score Form
Brown and Shavelson, *Assessing Hands-On Science: A Teacher's Guide to Performance Assessment.* Copyright 1996, Corwin Press, Inc.

How to Use Performance
Assessments in the Classroom

Overview

Performance assessments are useful classroom tools. When used peri-
odically throughout a science unit, they can focus attention on whether new
concepts and procedures are being understood. This chapter introduces just
such a performance assessment.

First, we will present an outline of a unit of instruction called Mystery
Powders (intended for upper elementary students, primarily fifth grade or
sixth grade, but could also be used at the middle school level). Then you will
experience an assessment that is *embedded* between the seventh and eighth
lesson of this unit. This assessment asks students to use what they have learned
about the physical and chemical properties of powders to identify what
powders are in a couple of mystery mixtures. Then you will learn to score some
students' assessment notebooks, using the scoring system that was developed
for this assessment.

Finally, this chapter will offer a way for you to summarize and then interpret your class's scores so that you can use the information to monitor your instruction.

Incidentally, we call this type of assessment *component identification* and with it we use an *evidence-based* scoring system.

Collect the materials you will need for Chapter 5.

MYSTERY POWDERS MATERIALS LIST

Embedded Assessment #2

- 6 small plastic cups (1 or 2 oz)
- Magnifying lens
- 6 wooden stir sticks
- 2 plastic spoons
- 1 *each* small containers of water, vinegar, and iodine (Luger's Concentrate, diluted 5 parts water to 1 part iodine, works best)
- 2 5 in × 7 in pieces of black construction paper
- 4 Baggies

Powders

4 teaspoons baking soda

9 teaspoons salt

2 teaspoons plaster of paris

5 teaspoons sugar

3 teaspoons cornstarch

Mix the following powders in separate unlabeled Baggies, shake, and set aside:

$\frac{1}{2}$ tsp baking soda with 4 tsp salt

$\frac{1}{2}$ tsp plaster of paris with 4 tsp salt

$\frac{1}{2}$ tsp baking soda with 4 tsp sugar

2 tsp cornstarch with 2 tsp baking soda

You will have about 1 teaspoon of each of the five powders left over. Place the remains of each powder in a separate little pile on the black construction paper for examination purposes.

Recall from the first chapter that Marilyn, a fifth-grade teacher, was conducting a science lab from a unit called Mystery Powders, in which students used their senses and some chemical tests to distinguish among five common household powders. During the lab, she walked around the room, helping students think through their use of chemical indicators such as iodine and vinegar to confirm or disconfirm the presence of a particular powder. Remember, Marilyn struggled with how she was going to assess what her students had learned; she wanted to know whether they could use the knowledge they had gained to gather both confirming and disconfirming evidence to reach a justified conclusion.

This chapter describes the assessment that Marilyn uses at this precise point in the Mystery Powders unit—an assessment that can inform students of their progress and teachers of how well their instruction is progressing and what might be done to improve it. We offer it here as an example of how performance assessments can be used in the classroom to meet teachers' testing needs.

Purposes of Classroom Testing

As a classroom teacher, why do you test your students in science?

Two reasons come to mind—both equally important. First, you are accountable for the learning that is taking place in your classroom; tests give you evidence of that learning, summed up in grades or scores that reflect student performance. A grade informs students about how their work measures up. In addition, when you conference with parents, a grade helps them know how their child is doing. Second, you as a teacher need to get a handle on the effectiveness of all those learning activities your students experience. You want to know whether students are catching on to particular concepts and procedures through whatever curriculum you are using. In essence, you test to evaluate student knowledge for your information and the students' and also to evaluate your instructional program.

Among the most useful kinds of performance assessments, therefore, are those that occur every so often throughout a unit. We call these embedded assessments because, to the students, they appear as a natural follow-up lesson to what has come before. To you as a teacher, however, embedded assessments offer a glimpse of where students are in their learning. They allow you to adjust the curriculum, go back and review points that were misunderstood, and help focus attention on the applicability of what has been learned. They also help students prepare for the end-of-unit assessment.

This chapter will demonstrate how embedded performance assessments can be used during the course of a unit. We will use as our example a unit called Mystery Powders, which is commonly taught at the upper elementary level.

The Mystery Powders Unit[1]

Students are given five ordinary white powders: baking soda, sugar, salt, cornstarch, and plaster of paris. Some teachers use the names of the powders; others simply call them A, B, C, D, and E. Over a period of weeks, the students gather evidence about each powder's physical and chemical properties, recording their findings in *laboratory notebooks*. As you can see from Figure 5.1, the unit is divided into nine activities of approximately one to two class periods each. During these lessons, students compare how the powders look under a hand lens, how they feel to the touch, and how they react to other substances, such as water, vinegar, and iodine. Students learn that each powder has a set of characteristics, both physical and chemical, that can be used to identify it, and that some characteristics are more informative than others for identification purposes. For example, they learn that iodine turns purple/black when mixed with cornstarch but not with the other powders. The iodine test, therefore, is a reliable way to confirm the presence of cornstarch. But sensory information, such as "powdery to the touch," is true of three of the powders, so the touch test is not adequate to identify cornstarch.

Mystery Powders Unit

SEQUENCE OF ACTIVITIES

1. Mystery powders introduction

2. What are the mystery powders?

3. Water test

4. Water test overnight

5. Vinegar test

6. Vinegar test overnight

7. Iodine test

8. Iodine test overnight

9. Heat test

Figure 5.1. The Mystery Powders Unit

Take a little time right now to examine the five *individual* powders you gathered at the beginning of this chapter. In particular, notice how they look under a hand lens, how they feel to the touch, and how they react to water, vinegar, and iodine. (Resource B includes a lab notebook, compiled by a student as she examined the powders, that provides answers to these questions.)

Goals of the Mystery Powders Unit

What are the goals of this unit? Well, we want students to learn about and observe the different physical and chemical properties of the common powders. We also want them to learn what it means to use observation to gather evidence to support or refute a hypothesis about what an unknown powder might be. They should learn that some evidence is more informative than other evidence and that both confirming and disconfirming evidence can be used to identify a powder. For example, students should learn that when you add vinegar to baking soda it fizzes, so adding vinegar can be used to both confirm the presence of baking soda (it fizzes) or disconfirm its presence (the powder being tested doesn't fizz). Thus, students learn not only about the characteristics of the powders but also what counts as *scientific* evidence.

How do we know whether students are catching on to these principles? As shown in Figure 5.2, after Lessons 3 and 7, short (15- to 30-minute) embedded performance assessments are used to test students' knowledge. What do these assessments look like? You are about to find out. The assessment you will be taking is given after Lesson 7 and is the second embedded assessment (Mystery Powders Embedded Assessment #2) the students experience. Notice that by this time, students have examined the powders by touch and under a hand lens and have observed how the powders react to water, vinegar, and iodine.

Experiencing Mystery Powders
Embedded Assessment #2

At the beginning of this chapter, you were given a list of materials and directions for mixing powders and placing them in four Baggies. To give you a feel for the mystery of Mystery Powders, pick two of the four Baggies at random—so you do not know which mixtures you have chosen. Then, label them bag *X* and bag *Y*. Use these two bags as you follow the directions on pages 67 and 68. You may also use any of the equipment you have gathered to help you. Keep in mind that you take notes on the first page of the assessment notebook and report your conclusions on the second page. Only the second page is scored. Do the assessment now and return to this page.

Scoring This Assessment

The mystery mixtures we use when we give this assessment to students are salt and baking soda in bag *X*, and salt and plaster of paris in bag *Y*. Our

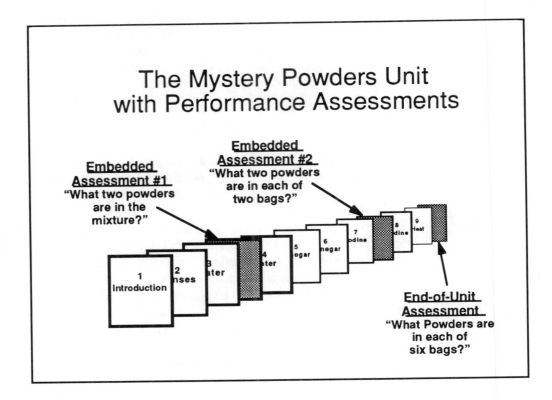

Figure 5.2. The Mystery Powders Unit With Performance Assessments
Brown and Shavelson, *Assessing Hands-On Science: A Teacher's Guide to Performance Assessment.*
Copyright 1996, Corwin Press, Inc.

scoring, therefore, will be based on those combinations of powders. Resource
C gives you a score form based on the other two combinations of powders in
case you randomly chose either the sugar and baking soda or the cornstarch
and baking soda and want to score your assessment later.

Figure 5.4 on page 70 is the score form that was designed to accompany
this assessment. Student performance is scored on three parts of page 2 of the
students' assessment notebook for each powder mixture. Notice that the first
column on the score form is "What's inside the bag?" This is the place to mark
if the student correctly identified the powders in the mixture. If both are
identified correctly, we give 1 point—otherwise, we give a 0. Next, look at the
third column—Test(s). Did the student name the tests used for identification?
Check off each test mentioned.

Look at the middle column and you will see that for each powder there is
a set of evidence. For baking soda, the best confirming evidence is that it fizzes
when vinegar is added. *Black boxes* are the best confirmatory evidence available. In the absence of this best confirmatory evidence, we can disconfirm the
presence of a powder by ruling out what it is not. Notice that for plaster of
paris, there are no black boxes,[2] but we know it is not cornstarch because it does
not turn purple/black when iodine is added, and we know it is not baking
soda because it does not fizz when vinegar is added. We also know that it is

My name   _____

My partner's name   _____

Find out what mystery powders are in bag X and what mystery powders are in bag Y. There are two powders in bag X and two powders in bag Y. Use any of the equipment on the table to help you determine what is in each bag.

HINT: You may have to do more than just one test.

<u>**Keep notes**</u> on what test(s) you did and what you observed in the space below. Use your notebook from science class to help you determine what is in each bag. When you think you know what is in a bag, record your results and conclusions in the table on page 2.

==

NOTES

Mystery Powder	Test I did	Observations

Figure 5.3. Mystery Powders Assessment #2 (1 of 2)

Brown and Shavelson, *Assessing Hands-On Science: A Teacher's Guide to Performance Assessment.* Copyright 1996, Corwin Press, Inc.

RESULTS AND CONCLUSIONS

Look at the tests and observations you made today.
Check your observations with the notebook you kept during
science class.
Fill in the table below.

Mystery Powder	What's inside the bag	What test told you	How did you know? What happened?

Figure 5.3. Mystery Powders Assessment #2 (2 of 2)
Brown and Shavelson, *Assessing Hands-On Science: A Teacher's Guide to Performance Assessment.* Copyright 1996,
Corwin Press, Inc.

not grainy—it is powdery—so we can rule out salt and sugar. By a process of elimination, we conclude that the mystery powder is plaster of paris. On this score form, disconfirming evidence is represented by *white boxes*. Other evidence may be mentioned, but it is merely contributory evidence and does not offer the best evidence for the presence of a powder. Now that you have some idea what to look for in each portion of the score form, we will look at a student's assessment notebook and score it.

In Figure 5.5, you will find page 2 of Jane's assessment notebook. Remember that we score only the second page of the notebook, because this is where Jane summarized her findings—gathering evidence to support her conclusions. We will talk you through the score form, and you can fill in the blank one on page 70 as you go. Then compare your score form with the completed one in Figure 5.6.

For powder *X*, Jane correctly identified "salt and baking soda," so on the score form in the first column we give her 1 point. If she had identified only one powder correctly (or none), we would give her 0 points. Now we look at the tests she used for powder *X*, noticing that she mentions only vinegar. We put a check in the third column of the score form beside vinegar. Incidentally, if a student puts an observation or test in the wrong column, we still count it as correct—as long as it is down in writing, it doesn't matter to us where it is placed.

Now we look at the observations Jane uses as evidence (middle column on the score form). She noticed "bubbes" that indicated the presence of baking powder, and "squar crystles" to indicate the cube-shaped crystals of salt. Checks are placed in both these ovals. Notice that we have made a judgment to accept "squar" instead of cube shaped. This seems like an easy decision to make—after all, they are almost synonymous words. Sometimes, however, the judgment is harder to make, and some consensus is needed about what synonyms will or will not be accepted.

Next, we will figure out what score Jane is given for the evidence she gathered for powder *X*. Look at the Quality of Evidence box at the bottom of the page. All the black boxes are checked (there were no white boxes for powder *X*), so she gets 4 points. However, read the note at the very bottom of the point box. Jane loses ½ point because she forgot to mention one test, but we don't subtract any more than ½ point, even if she hadn't named *any* tests. So her Quality of Evidence score is 3.5.

Now we'll score powder *Y*. Jane correctly identified "salt" and "plaster," so she is given 1 point. She did not mention any tests, so no checks are given under Tests. Under Observations, we credit her for one piece of disconfirming evidence, "no fizz," and the confirming evidence for salt, "square crystals." What is her Quality of Evidence score? We can't give her 4 points, because she did not mention all the disconfirming evidence (white boxes); but she did get one black box and one white box, so 3 is her score so far. Now we need to deduct ½ point because she forgot to write down the tests that she used. So her Quality of Evidence score for powder *Y* is 2.5. Now, simply add the totals for Correct Answer and Quality of Evidence Scores.

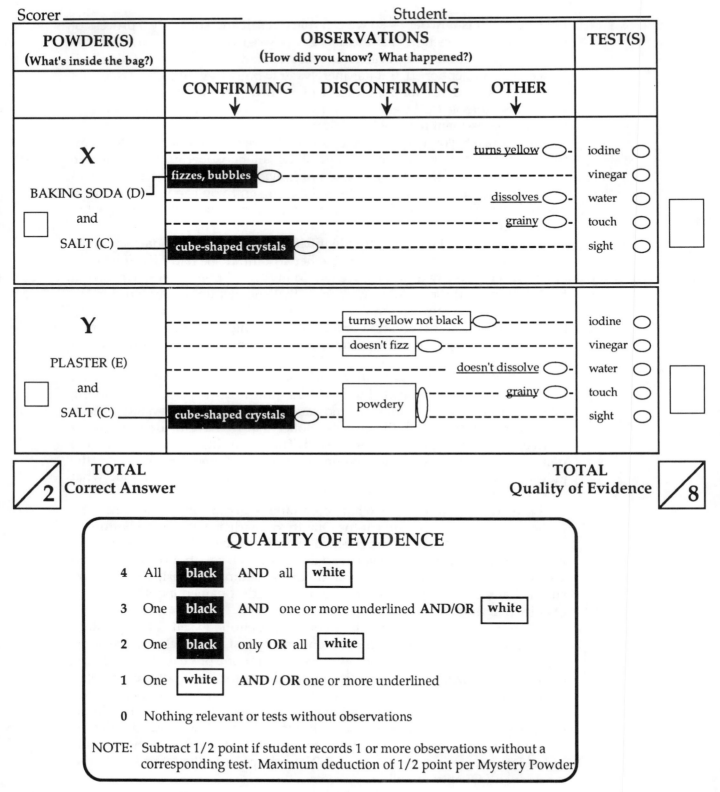

Figure 5.4. Mystery Powders Embedded Assessment #2 Score Form

Brown and Shavelson, *Assessing Hands-On Science: A Teacher's Guide to Performance Assessment.* Copyright 1996, Corwin Press, Inc.

RESULTS AND CONCLUSIONS

Look at the tests and observations you made today.
Check your observations with the notebook you kept during science class.
Fill in the table below.

Mystery Powder	What's inside the bag	What test told you	How did you know? What happened?
X	Salt, baking soda	Vinegar	bubbes, squar Cryshes
Y	Salt, Plaster	no fizz	squar Cryshes

Figure 5.5. Jane's Assessment Notebook

Brown and Shavelson, *Assessing Hands-On Science: A Teacher's Guide to Performance Assessment.* Copyright 1996, Corwin Press, Inc.

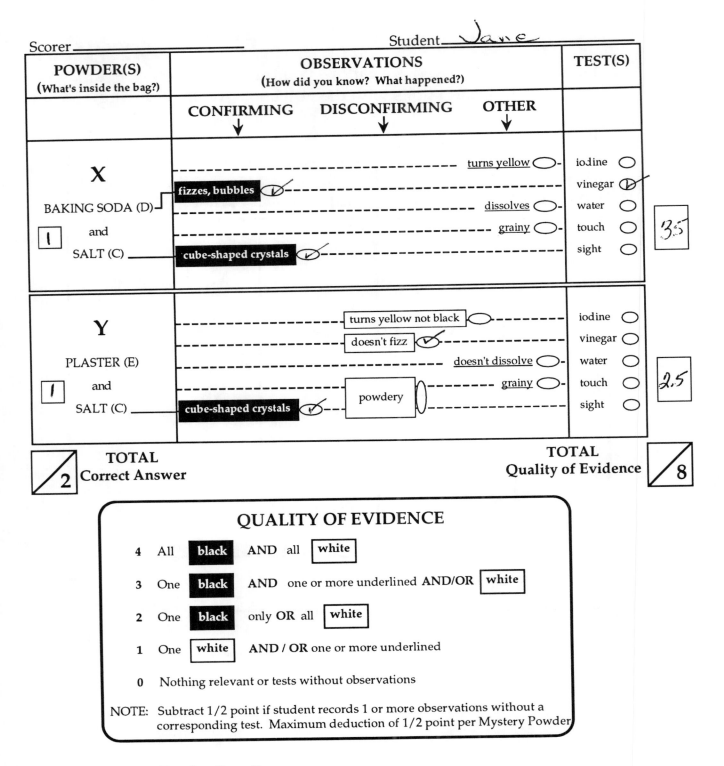

Figure 5.6. Mystery Powders Score Form
Brown and Shavelson, *Assessing Hands-On Science: A Teacher's Guide to Performance Assessment.* Copyright 1996, Corwin Press, Inc.

Scoring Sam's Assessment Notebook

On page 75 there is the second page of another student's assessment notebook for you to practice scoring with. You will find a blank score form on page 74. Go ahead and score it now.

When you're done, check your scores with ours on page 76 (Figure 5.9) to see if they agree. Remember that you are not expected to have mastered the score form in such a short time; however, we have found that by the time teachers score about 10 notebooks, they are comfortable with the score form.

Here is our rationale for Sam's scores, just in case you have questions.

- **Powder X**

Correct Answer Score = 1

Baking soda and salt correctly identified.

Quality of Evidence Score = 3

One black box checked and one underlined statement (turns yellow), with both tests named. Scorer opted not to count "nothing happ[en]ed" as meaning the same as "dissolves." Score would be the same in any case.

- **Powder Y**

Correct Answer Score = 0

Reported salt and sugar instead of plaster and salt.

Quality of Evidence Score = 1

Found two pieces of disconfirming evidence (turned yellow with iodine, and didn't fizz with vinegar).

- **Sam's Total Scores**

Total Correct Answers = 1
Total Quality of Evidence = 4

Interpreting Scores

Remember that one of the main reasons we are testing students is to monitor the effectiveness of instruction. How can a teacher tell from looking at students' assessments exactly which concepts and procedures are understood and which are not? In fact, this is where giving this assessment really pays off—here's a strategy for interpreting scores with direct implications for teaching.

Using Marilyn again as our example, let's assume that she has just finished testing her students with this assessment. She collects the assessment notebooks and *scores* them with the score form used above. Then she selects a sample of students' performances by picking 10 students' score forms at random. These 10 will give her an idea of how well her class, in general, understands the Mystery Powders unit so far.

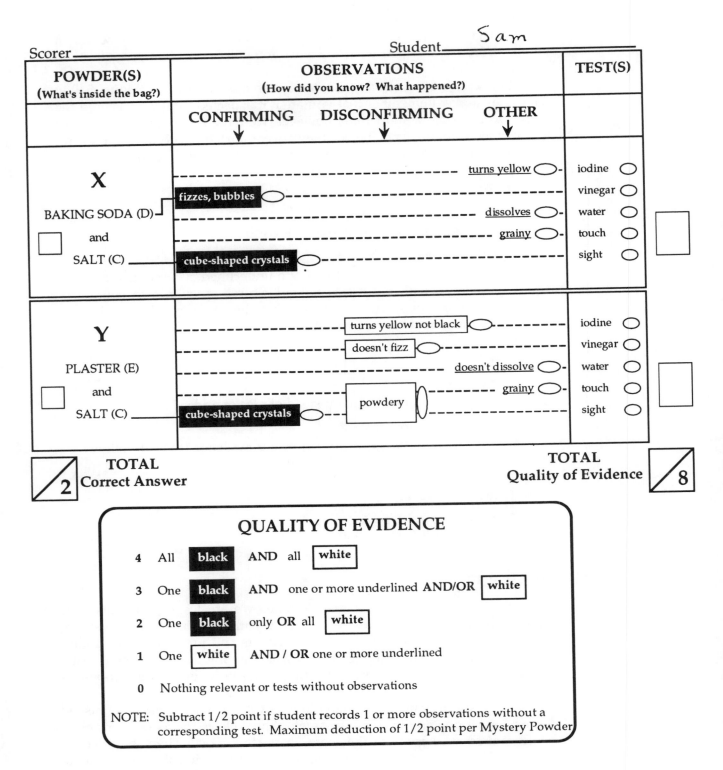

Figure 5.7. Mystery Powders Score Form
Brown and Shavelson, *Assessing Hands-On Science: A Teacher's Guide to Performance Assessment.* Copyright 1996, Corwin Press, Inc.

RESULTS AND CONCLUSIONS

Look at the tests and observations you made today.
Check your observations with the notebook you kept during
science class.
Fill in the table below.

Mystery Powder	What's inside the bag	What test told you	How did you know? What happened?
X	Salt & Baking soda	Smell Ioddin vineger touch	With Ioddin it trund yellow with watr nothing happed with vineger it fizzd up
Y	Salt & suger	Smell Ioddin vineger touch	With Ioddin it trund yellow with vineger it didn't fizz

Figure 5.8. Sam's Assessment Notebook

Brown and Shavelson, *Assessing Hands-On Science: A Teacher's Guide to Performance Assessment.* Copyright 1996, Corwin Press, Inc.

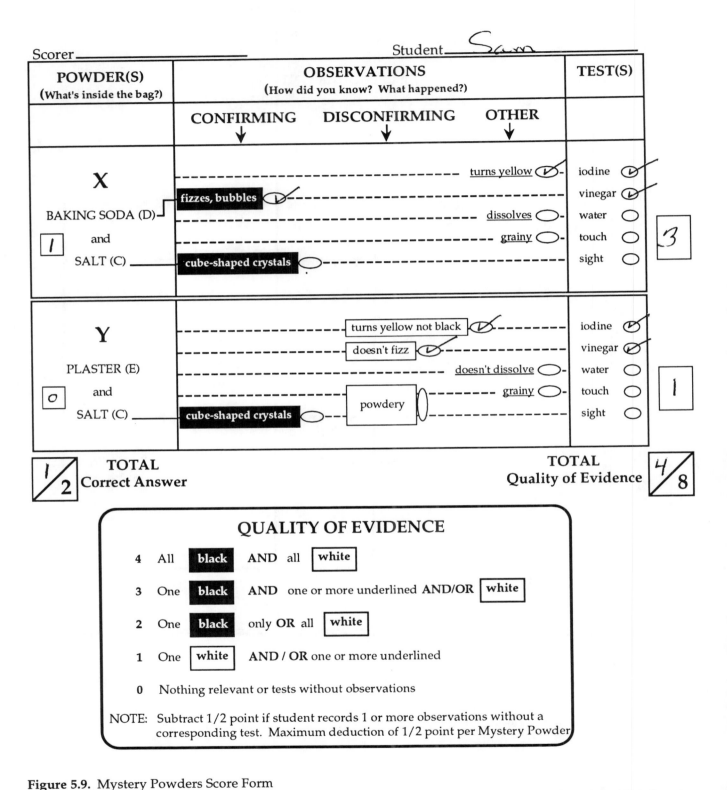

Figure 5.9. Mystery Powders Score Form

Brown and Shavelson, *Assessing Hands-On Science: A Teacher's Guide to Performance Assessment.* Copyright 1996, Corwin Press, Inc.

Marilyn *compiles the scores* into a table so she can easily find patterns that arise. First, she places the 10 score forms in order according to their Quality of Evidence score, highest score on top. Then she fills out the Score Summary sheet you see on page 78. For each student, she fills in the Total Correct Answers score, Total Quality of Evidence score, number of black boxes used (confirming evidence), number of white boxes used (disconfirming evidence), if other evidence was used, and if any test was mentioned without a corresponding observation. Then, if students lost ½ point because they forgot to write down all the tests they used, an asterisk (*) is placed next to the Quality of Evidence score.

Once Marilyn has all the information entered for each of the 10 students, she looks at her Score Summary Form to see if any patterns arise; that is, she *interprets* those patterns. Figure 5.11 shows a sample taken from a typical class, like Marilyn's.

How well do the students understand confirming evidence? Can students use disconfirming evidence when confirming evidence is not available? Do they understand the limitations of other evidence? (We wouldn't want them to use exclusively this less useful type of evidence.) How consistent are they at recording both observations and tests used as evidence? All this information can help Marilyn *make decisions* in planning subsequent lessons.

Think and Respond

Put yourself in Marilyn's place. Based on what you see in Figure 5.11, how would you interpret the patterns of scores in this class and what decisions would you make concerning future instruction?

_____

_____

_____

_____

_____

_____

_____

_____

When we asked other teachers to consider the patterns they saw in this class and the implications of those patterns, they gave us the following responses:

1. All but two students used some confirming evidence. It seemed that almost everyone understood this concept to some extent; however, only two students found all possible pieces of confirming evidence. (When we looked back at the score forms, we found that the confirming evidence for salt was especially difficult for the students.)

INSTRUCTIONS FOR USING THE SCORE SUMMARY

Score all the assessments. Choose 10 students at random from the entire class. Look at the Total Quality of Evidence score (see box at lower right of score form) and put the students in order from the highest to lowest. Fill in the following table.

Column I   Record the Total Correct Answers score from the box at the lower left of the score form.

Column II   Record the Total Quality of Evidence score from the box at the lower right of the score form. Put an * next to the score if the student received half points for one or more of the Mystery Powders.

Column III   Count and record the number of times confirming evidence was used.

Column IV   Count and record the number of times disconfirming evidence was used.

Column V   Put a check if the student used one or more types of evidence. They are the underlined observations.

Column VI   Put a check mark if one or more tests are recorded without the corresponding observations.

ANSWER SUPPORTED BY EVIDENCE		QUALITY OF EVIDENCE			
I	II	III	IV	V	VI
Total Correct Answers (0, 1 or 2)	Total Quality of Evidence (0 to 8)	Confirming (# black boxes) (0 to 3)	Disconfirming (# of white boxes) (0 to 3)	Other (put check if present)	Tests without Observations
Highest					
Lowest					

Figure 5.10. Instructions for Using the Score Summary

Brown and Shavelson, *Assessing Hands-On Science: A Teacher's Guide to Performance Assessment*. Copyright 1996, Corwin Press, Inc.

2. Only four students used disconfirming evidence, and this was minimally used (each used only one observation and corresponding test). The teacher needs to help students further develop their use of disconfirming evidence.

3. Three students reported other evidence. If students are reporting only this weaker form of evidence and not reporting confirming, disconfirming evidence, or both, it might suggest they do not understand the relative importance of each type of evidence. Perhaps the teacher needs to focus more on these issues.

4. Students reported observations without the corresponding tests (those scores with an asterisk beside them). Students might not be aware of the importance of both observations and tests as *evidence*. The teacher needs to emphasize proper recording of observations and tests.

Here's the interpretation strategy in a nutshell:

> *Score* student performance
>
> *Compile a table*, using a sample of scores
>
> *Interpret* patterns in the scores
>
> *Make decisions* based on those patterns

ANSWER SUPPORTED BY EVIDENCE		QUALITY OF EVIDENCE			
I	II	III	IV	V	VI
Total Correct Answers (0, 1 or 2)	Total Quality of Evidence (0 to 8)	Confirming (# black boxes) (0 to 3)	Disconfirming (# of white boxes) (0 to 3)	Other (put check if present)	Tests without Observations
1	7	3	1	√	
2	6*	3	1		
2	5*	2	1	√	
0	4	2	0		
1	3	1	1		
2	3	1	1		
0	2	1	0		
0	1.5*	0	0	√	√
0	1.5*	1	0		
0	0	0	0		

Highest (aligned to row with 1, 7, 3, 1, √)
Lowest (aligned to bottom row with 0, 0, 0, 0)

Figure 5.11. Sample Score Summary
Brown and Shavelson, *Assessing Hands-On Science: A Teacher's Guide to Performance Assessment.* Copyright 1996, Corwin Press, Inc.

Linking Assessment With Curriculum

An embedded performance assessment goes hand in hand with your curriculum. To the student, it looks just like another learning activity; but for you as a teacher, it provides a reliable standard of measurement by which you can evaluate student performance and inform your teaching. Just remember, during the assessment your role is *observer*—not *teacher*. It may be very tempting to help students along during the assessment—but *resist that urge*. The assessment's value is in telling you what students can do *on their own*.

Embedded assessments send a message to students about what they are expected to know and be able to do. In this way, the assessment helps clarify and focus attention on the goals of the unit for both students and teacher, making the purpose of the unit more concrete.

They also enable you to test students' skills progressively, moving from relatively easy investigations at the beginning of a unit to more difficult ones, culminating with the end-of-unit assessment.

The End-of-Unit Assessment

At the end of a unit, another performance assessment is given to provide comprehensive information about what the students have learned over the whole unit. This assessment can offer a dependable method both for assigning grades to report student progress and for planning future instruction. Many teachers have told us how valuable the end-of-unit assessment has been to decisions about how a particular unit will be taught the following year.

Summary

This chapter described a widely used unit of instruction, called Mystery Powders, to demonstrate how performance assessments can be embedded within the lessons of a unit to inform teaching. First, you did the performance assessment yourself; then you scored some student assessment notebooks, using another type of analytic score form (called an evidence-based score form), which allowed you to evaluate both the correct answers and the quality of the evidence students used to justify their conclusions.

Next, we presented a way to interpret scores from the embedded assessment, using a sample from a whole class. By completing a Score Summary sheet you were able to tease out direct information about the procedural and conceptual knowledge of students—information that had immediate implications for future lessons.

Both embedded and end-of-unit performance assessments can give the teacher and the students valuable information about student knowledge in science. We will next look at another kind of assessment and score form. This one was developed as part of the pilot test of a performance assessment in a statewide science achievement test. It is scored with a holistic *scoring rubric*, a scoring system that particularly lends itself to large-scale testing situations.

Notes

1. This unit was developed by the Educational Development Center. It is now available through the Optical Data School Media. We include their address and phone number in Resource D.

2. The best confirming evidence for plaster of paris is that it hardens with water after a period of time, depending on the dilution. But in the 15 to 30 minutes it takes to do this assessment, that test is not an option, so we depend on disconfirming evidence instead.

What Is a
Holistic Scoring System?

Overview

You have probably heard about scoring *rubrics* and wondered where they fit into this "scoring game."

This chapter introduces an assessment that was part of the pilot study for a statewide performance assessment in science given at the sixth-grade level. It is scored with a holistic scoring rubric. You will do the assessment and then score students' assessments using the rubric. We'll close the chapter with a discussion of the advantages and disadvantages of analytic versus holistic scoring of performance assessments. But first gather the materials you will need for this chapter (see page 83).

An Example From a Statewide Performance Assessment

Picture a classroom full of students taking a statewide achievement test in science. But in addition to using the usual test booklet, #2 pencil, and fill-in-the-bubble test form, students are engaged in doing hands-on investigations. This alternative kind of performance-based large-scale testing is being developed and piloted by states across the country in hopes of better assessing students' science knowledge. The assessment you are going to do in this chapter was originally part of a pilot test for this kind of statewide assessment.

LEAVES MATERIALS LIST

Performance Assessment
(about 6th grade)

- 10 different green leaves, artificial or real (look for variety in size, shape, and features)

- 2 small Baggies, labeled *Bag A* and *Bag B*

Directions to prepare materials:

Number each leaf on the back (1 to 10) with a pen

Places leaves #1-9 in *Bag A*

Place leaf #10 in *Bag B*

You will use a scoring system that differs dramatically from the kinds we have used in the past two chapters. With Paper Towels and Mystery Powders, you used what we call an *analytic* scoring system, which helps you *analyze* the components of a student's performance—that is, helps you come up with a score or grade to evaluate both the answer and the way the answer was reached. But there is another popular kind of scoring system that you might have heard a lot about. This is called a *scoring rubric,* and it differs from the analytic systems in that it is *holistic* in nature, giving an overall judgment of student performance.

We are going to be examining the advantages and disadvantages of rubric scoring in this chapter. First, however, we'll do another performance assessment—one that asks students to categorize.

Experiencing the Leaves Assessment

As we mentioned above, this assessment, called Leaves,[1] was originally part of California's pilot study for a statewide achievement test in science given to sixth-grade students. To take this test, students rotated through five "stations," each with a different set of materials and a task. You are going to do just one of the tasks in this assessment. When this test was given, students worked in pairs, keeping individual written booklets. You can work alone or with a partner.

Here are your directions based on those written in the administration guide:

- This is a science test in which you will get to do an experiment.
- Please listen carefully to the directions.
- This test is probably a little different from the tests you are used to taking. I hope you will enjoy it. This may be a task that you have not had a chance to do before. Even if you are not sure what to do, I want you to try the task.

- You will be working in pairs, but each of you will record work individually in your own assessment booklet.
- You will be given 13 minutes to complete the task. After 11 minutes, you will be told you have 2 minutes remaining to complete the task. You will be told when you have 30 seconds left.
- If you finish before time is up, carefully check your work. After 13 minutes, you will be told to stop.
- Please leave your station the way you found it.
- You should have two bags (*A* and *B*). Bag *A* should contain 9 leaves; Bag *B* should contain 1 leaf. Check to see that you have all the materials you need. (Wait a moment for them to do this.)
- Please raise your hand if you are missing materials. (Supply missing materials.)
- Are there any questions?
- The task (remember, you are doing only one) has a two-page answer sheet.
- Keep working on your answer sheet for this task until you reach the stop sign at the end.
- We are now ready to begin the test.
- Silently read the instructions and begin the task NOW (you will find it on pages 85-86), and then return to this page.

Components of the Leaves Assessment

Before we go on to scoring, let's review what we know so far about this assessment. It included a classification task (divided into subtasks) and a response format. Your task was to set up a classification system for the leaves in bag *A* and describe the properties of the system by which you grouped them. Then you were to consider the leaf in bag *B*, decide if it fit in with your original category system, and then change the system (regroup the leaves), if you thought it was necessary, to incorporate the new leaf. Your last subtask was to explain why or why not you regrouped when the new leaf was added. Notice that the assessment gives you pretty clear step-by-step directions for carrying out the investigation. We think you will agree that it is a highly structured task. The response format is also highly structured in that you are provided two tables with labels—all you have to do is fill them in when you have grouped your leaves.

By providing this much structure to the task and response format, the test developers set strict limits on the amount of information they gain about students' ability to classify and display findings. For example, although the assessment tells us if a student understands the concept of *grouping* by leaves' attributes, it does not tell us if the student can create a table to classify leaves.

We have examined the assessment's task and response format, but what about its scoring system? Like other large-scale performance tests, it uses a scoring rubric that was created to score the task holistically. Just what is a scoring rubric?

California Assessment Project (CAP)

TASK 2   LEAVES

Directions: You have collected some leaves. Open the bag marked "A" and spread the leaves on the table. Observe each leaf and group the leaves according to their characteristics. You may make as many groups as you like.

1. Look at the groups that you made. Record the number of each leaf in your group(s) on the chart below. Describe how the leaves within each group are alike.

Group - Leaf Numbers	Description of Groups

2. Open the bag marked "B". Take out the new leaf and observe its characteristics. Compare it to the groups you just made. Into which group would you put this new leaf? Why?

_____

_____

_____

GO ➡

Figure 6.1. Leaves Performance Assessment (1 of 2)

© California State Department of Education, Bill Honig, Superintendent of Public Instruction, Sacramento, 1990.

California Assessment Project (CAP)

TASK 2	LEAVES

Directions: Scientists do not always agree on how things should be classified. Put your leaves back together adding the new leaf. Observe the characteristics of the leaves. Put the leaves into groups.

3. Look at the groups that you made. Record the number of each leaf in your group(s) on the chart below. Describe how the leaves within each group are alike.

Group - Leaf Numbers	Description of Groups

4. Did the new leaf from Bag B change how you grouped the leaves? If it **did**, explain why. If it **did not**, explain why not.

_____

_____

_____

Please put leaves numbered 1 - 9 into "BAG A"
Please put leaf numbered 10 into "BAG B"

Figure 6.1. Leaves Performance Assessment (2 of 2)
© California State Department of Education, Bill Honig, Superintendent of Public Instruction, Sacramento, 1990.

Using a Holistic Scoring Rubric

A *holistic scoring rubric* is a set of statements that describe various levels of student performance, ranging from poor to outstanding. It may also include examples of student performance at each level. The following box contains the rubric created for Leaves. Scorers take into consideration the student's response to the whole task and judge it according to how it fits the description. Notice that the rubric is designed to allow for a wide range of student responses, and to help scorers, the statements are often accompanied by examples of student work at each level.

LEAVES RUBRIC

Outstanding: Rating = 4
Gives complete answers with logical responses; uses complex attributes (beyond color and size); development of a process of thinking is evident; provides both detailed and specific comparisons; rationale is clearly stated and consistent with groupings; demonstrates understanding of the concept; may include pictures in addition to written explanation; descriptions match object; completes task.

Competent: Rating = 3
Gives complete answers with some logical responses; shows development of a process of thinking; sortings are clearly stated; descriptions are clear—may have only basic attributes; rationale is consistent with groupings; shows understanding of the concept; second page completed, however, only the first page is outstanding.

Satisfactory: Rating = 2
Gives answers that are simplistic or incomplete; explanations (rationale) are not consistent with results; some leaves are not accounted for; shows basic understanding of the process of grouping; task not necessarily completed on the second page.

Serious Flaws: Rating = 1
Gives answers that are not complete or understandable; doesn't show understanding of the process of grouping (no grouping or illogical groupings); may group some leaves but doesn't provide explanation or rationale, questions #2 and #4 are not answered or are answered inaccurately; task is not completed.

No Attempt: Rating = 0

Scoring Student Notebooks

Trying out a rubric is the best way to get a handle on it, so first we'll examine Jacob's notebook, found on pages 89 and 90. Go ahead and read through his notebook and decide where you think his response fits on the Leaves rubric. What score would you give him and why?

Other scorers rating Jacob gave the following rationale for their scores:

- Jacob completed the entire investigation, therefore he rates above a 1 or 2.
- Jacob gives complete answers with logical responses for all questions. This would put him at a 4 rating.
- His sortings are detailed and clearly explained.
- Both pages are outstanding. To rate a 3, only the first page need be outstanding; but with both pages outstanding, he earns a 4.

In scoring this assessment, three trained scorers gave Jacob a 4.

Now read Leticia's assessment notebook, on pages 91 and 92, and decide where her performance falls on the rubric.

What score did you give Leticia? On what did you base your score? Here's what other scorers gave as their rationale:

- Leticia did not give complete answers for questions #2 and #4. She did not answer "why" for either question. This would eliminate a rating of 3 or 4 and put her at a rating of 2 or 1.
- For questions #1 and #3, she drew pictures rather than give a written explanation. This is a satisfactory response.
- Although her descriptions of the groups in questions #1 and #3 are simplistic, they are complete, so this would be a difference between a 1 or 2 rating.
- For a rating of 1, questions #2 and #4 are either not answered or are answered inaccurately. Although Leticia did answer the first part of these questions, she did not answer the second part, "why?"
- When all of these items are considered as a whole, we consider her rating a 2.

In scoring this assessment, three trained scorers gave Leticia 2, 1, and 2. Obviously, some disagreement about scores is to be expected; however, with a well-designed rubric and training, scorers have been found to be highly consistent in their use of rubrics.

Comparing Scoring Systems

In this handbook, we have now used three kinds of scoring systems; two were *analytic* (they analyzed the component parts of performance), and one was *holistic* (it looked at performance as a whole). We can divide analytic scoring systems further: One—that used for Paper Towels—focused on the procedures students used (so we call this a procedure-based scoring system),

California Assessment Project (CAP)

TASK 2	LEAVES

Directions: You have collected some leaves. Open the bag marked "A" and spread the leaves on the table. Observe each leaf and group the leaves according to their characteristics. You may make as many groups as you like.

1. Look at the groups that you made. Record the number of each leaf in your group(s) on the chart below. Describe how the leaves within each group are alike.

Group - Leaf Numbers	Description of Groups
#'s, 1 ; 9 ; 6 ; 2 : Group B	they all have 3 points they all both sort of like a hat.
#3 Group L	it has no points and has many curved sides.
#'s, 5 ; 8 ; 4 ; 7 ; Group T	they all have two side which come together with a point at the top. They are mostly green

2. Open the bag marked "B". Take out the new leaf and observe its characteristics. Compare it to the groups you just made. Into which group would you put this new leaf? Why?

into the group that has numbers 5 ; 8 ; 4 ; 7 ; because it also has two side which came to a point. It is mostly green.

GO ➡

Figure 6.2. Jacob's Notebook (1 of 2)
© California State Department of Education, Bill Honig, Superintendent of Public Instruction, Sacramento, 1990.

California Assessment Project (CAP)

TASK 2   LEAVES

Directions: Scientists do not always agree on how things should be classified. Put your leaves back together adding the new leaf. Observe the characteristics of the leaves. Put the leaves into groups.

3. Look at the groups that you made. Record the number of each leaf in your group(s) on the chart below. Describe how the leaves within each group are alike.

Group - Leaf Numbers	Description of Groups
#'s 5; 4; 7; 8; 10;	they all have two sides that came to a point and they are mostly green.
#3	it has many round edges with no points
#'s 2; 6; 9; 1;	they all have three points.

4. Did the new leaf from Bag B change how you grouped the leaves? If it **did**, explain why. If it **did not**, explain why not.

no because it fit with group #'s 5; 4; 7; 8; 10; but it had jagged edges.

Please put leaves numbered 1 - 9 into "BAG A"
Please put leaf numbered 10 into "BAG B"   (STOP)

Figure 6.2. Jacob's Notebook (2 of 2)
© California State Department of Education, Bill Honig, Superintendent of Public Instruction, Sacramento, 1990.

California Assessment Project (CAP)

TASK 2   LEAVES

Directions: You have collected some leaves. Open the bag marked "A" and spread the leaves on the table. Observe each leaf and group the leaves according to their characteristics. You may make as many groups as you like.

1. Look at the groups that you made. Record the number of each leaf in your group(s) on the chart below. Describe how the leaves within each group are alike.

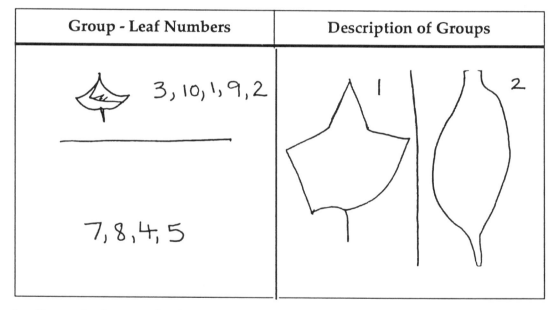

Group - Leaf Numbers	Description of Groups
3, 10, 1, 9, 2	
7, 8, 4, 5	

2. Open the bag marked "B". Take out the new leaf and observe its characteristics. Compare it to the groups you just made. Into which group would you put this new leaf? Why?

    It would go with my group two.

GO ➡

Figure 6.3. Leticia's Notebook (1 of 2)
© California State Department of Education, Bill Honig, Superintendent of Public Instruction, Sacramento, 1990.

California Assessment Project (CAP)

TASK 2	LEAVES

Directions: Scientists do not always agree on how things should be classified. Put your leaves back together adding the new leaf. Observe the characteristics of the leaves. Put the leaves into groups.

3. Look at the groups that you made. Record the number of each leaf in your group(s) on the chart below. Describe how the leaves within each group are alike.

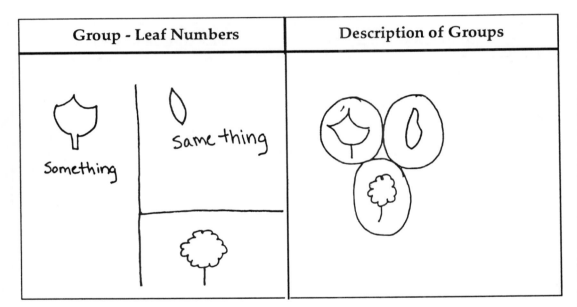

Group - Leaf Numbers	Description of Groups

4. Did the new leaf from Bag B change how you grouped the leaves? If it **did**, explain why. If it **did not**, explain why not.

    It made 3 groups

Please put leaves numbered 1 - 9 into "BAG A"
Please put leaf numbered 10 into "BAG B"

Figure 6.3. Leticia's Notebook (2 of 2)
© California State Department of Education, Bill Honig, Superintendent of Public Instruction, Sacramento, 1990.

and the other—that for Mystery Powders—focused on the quality of evidence the students gathered (so we call this an evidence-based scoring system).

These analytic scoring systems look at student performance as distinct but closely related pieces. The score form itself allows us to judge both the parts of the performance and how they combine to make for a successful investigation. For example, with the Mystery Powders score form (page 70), we were able to tease out a student's use of confirming evidence and disconfirming evidence, thereby helping us see how an individual student gathered evidence to support his or her conclusions. This turned out to be useful information when interpreting how the class as a whole understood the concepts and procedures of the unit; it was valuable for informing instruction.

Similarly, for Paper Towels (page 43), we can look at the score form for a student and know which parts of his or her procedures contributed to a successful or unsuccessful investigation. Knowing this aids us in helping the student.

With a scoring rubric, however, we have a general description of what performance is like at a particular level. We do not know exactly what it was about the performance that placed it at that level. For example, when we decided that Leticia's performance was a 2, it was the combination of factors about her assessment that led us to that decision, and it would be difficult for anyone to look at that 2 and decide specifically how to help Leticia. Holistic scoring is quick and reliable with trained scorers, but it does not give the kind of useful detail about performance that an analytic scoring system does.

What conclusions can we draw from a comparison between scoring systems? The type of scoring system used needs to be determined by the purpose for testing. If you want to use the scores to inform classroom instruction, analytic scoring systems give you more useful information. If you want a quick, overall picture of a student's performance, say for large-scale accountability purposes or classroom accountability, a rubric may be just what you are looking for.

Summary

This chapter introduced a new assessment, Leaves, which was part of the pilot for a statewide achievement test. It was scored holistically, with a scoring rubric.

Well-defined rubrics are quick and reliable to use (by trained scorers), but they do not provide as much useful information to inform instruction as do analytic score forms, such as procedure-based or evidence-based score forms. Large-scale tests, however, are usually scored with holistic rubrics because they take less time to use and can produce consistent scores across different scorers.

Note

1. This task, copyrighted by the California State Department of Education, is used with its permission.

Choosing an Assessment
You Can Trust

Overview

Are all performance assessments created equal? How can you choose a performance assessment that gives you information you can rely on about what your students know and can do in science?

This chapter gives you some practical hints for choosing trustworthy performance assessments. We suggest you consider four important qualities when evaluating the appropriateness and worth of a particular performance assessment for use in your classroom. These qualities are *reliability, validity, utility,* and *practicality.*

Issues of *reliability* concern the consistency of a student's scores. A student's score should be consistent no matter who scores the test or what particular relevant task the student is asked to perform. This chapter will present our research findings concerning the reliability of performance assessments, encouraging you to carefully examine potential assessments for their reliability.

Performance assessments also need to have *validity* for the purpose for which they are used. In order for them to be *content valid*, students need to have had an opportunity to learn both the content and the procedures being assessed. Another validity issue presented in this chapter is the *exchangeability* of testing methods. Does an assessment done on a computer, for example, measure the same type of knowledge and skills as the same assessment done with "real materials"? What about the exchangeability of paper-and-pencil performance tests with hands-on assessments? Exchangeability refers to these and similar questions.

The *utility* of a performance assessment for evaluating students and monitoring instruction is another important quality to consider when selecting a particular assessment. We will give you some questions to ask about the utility of assessments you are considering using.

And last, but certainly not least, we ask you to consider how practical the performance assessment is. Issues of *practicality*, such as cost and ease of administration, are vitally important to this new testing technology. In fact, this issue may be foremost in many teachers' minds when thinking of alternative assessments. We agree; teachers need to be realistic and confront the practicality of performance assessments.

Critically Testing the "Test"

Tests are powerful tools. We give them power when we interpret their results as representing student knowledge. Even when high stakes decisions are not being made with the results, the scores carry power in the classroom—just by their very nature. But is our trust in them well founded?

A performance assessment, like any other test, needs to be scrutinized as to its *trustworthiness*. After all, if you can't trust it to give you consistent, unbiased, and valid results, it is not worth the effort of using it. In fact, you would be fooling yourself about the information you have gleaned from all that time and work.

On the other hand, a good (trustworthy) test can be a tremendous help to you as a teacher. It can efficiently produce a measurement of your students' knowledge and/or ability in some area. You can then use this information, along with other indicators (your own observation, projects, homework, etc.), to inform instruction and document student knowledge. But if you can't trust your testing instrument, you don't know what you have.

It probably won't surprise you that testing experts are extremely zealous about the trustworthiness—called the reliability and validity—of tests. But teachers also need to be knowledgeable about these issues, especially as they relate to performance assessments, because as new performance assessments become available, issues of trustworthiness will undoubtedly become clouded.

Trustworthiness alone is not enough to make a test worthwhile; it must also be useful and practical. In this chapter we will present to you four qualities to look for when you are deciding whether to use a particular performance assessment. They are:

- Reliability
- Validity
- Utility
- Practicality

Of course, before you can even apply these qualities, you need to decide whether the assessment you are considering really is a performance assessment. Does it consist of a task, a response format, and a scoring system? Remember, if you do not have *all three parts,* you do not have a performance assessment.

But let's assume that the performance assessment you are considering does meet this definition. What else should you look for?

Reliability

Reliability is simply another word for consistency. We would never use a stretchy piece of elastic as a tape measure, because it would give inconsistent—unreliable—results. In the same way, we want an assessment instrument that does not "stretch" but is reliable. A student's score should remain relatively consistent no matter who scores the test (called interscorer or interrater reliability) and no matter what particular relevant task we ask the student to do to demonstrate knowledge (called intertask reliability). Let's take these one at a time.

Interrater Reliability

Remember when you scored Paper Towels without a standardized score form, in Chapter 3? The way you scored performance might have been very different from the way another teacher scored it. You might have used different criteria in your judgment and given different weight to various aspects of the performance. Or you might have used letter grades, and another teacher might have used a 6-point scale. But when you scored performance using the Paper Towels score form, you knew what to look for in the investigation; you had a guide to the criteria used to judge performance. Moreover, you had a fixed grading system on which to order performance. Another teacher trained to use that score form could come up with virtually the same score (grade) you did. In fact, this is exactly what happened when we trained scorers to score more than 200 students doing the Paper Towels assessment. We found very *high* interrater reliability.

What do we mean by high? Reliability is indexed numerically by using a special kind of *correlation coefficient* called a *reliability coefficient*. The coefficient ranges between 0 (indicating no reliability) and 1 (perfect reliability). This special correlation simply indicates the extent to which two raters score a group of students' performances consistently, in numerical form.

You can get a picture of what interrater reliability looks like on a scatter plot (Figure 7.1) by looking at the scores given by one teacher (Rater 1) and those given by another teacher (Rater 2) on the same group of 10 student assessment notebooks.

Each ■ represents the scores given to one student by two raters. (You read this plot by looking across and down from a particular ■ to the scores given by Rater 1 and Rater 2.) Notice that the scores with boxes around them have

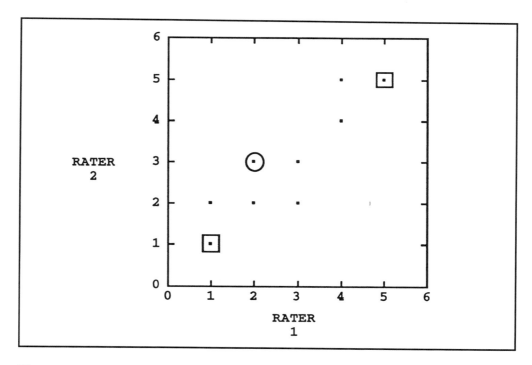

Figure 7.1. Interrater Reliability

perfect consistency—both raters gave 5s and 1s to the same students. The score with a circle around it is not as consistent—close but not exactly the same. Rater 1 gave the student a 2, and Rater 2 gave the same student a 3. Generally speaking, the ■s shown on this graph suggest that the two raters were fairly consistent; that is, if one assigned a low score to a student, the other also tended to assign a low score, and so on.

As mentioned before, reliability can be indexed quantitatively. The plot in Figure 7.1 has a reliability coefficient of .91, which is very high.

Figure 7.2 shows you what various levels of reliability look like on scatter plots. You can see that if the reliability is much below .80, one scorer might be giving a student a low score while the other scorer is giving him or her a high score. Perhaps scorers are judging the same performance by somewhat different criteria—definitely not consistently.

Remember, we said the interrater reliability for the Paper Towels assessment was very high; in fact, the correlation in our research was .94.

Even though interrater reliability is very important, developers of performance assessments might not include the reliability coefficients with their tests. If they are not included, we suggest you ask the developers about the interrater reliability of their scoring systems, and be sure the coefficient is *above .80.*

Intertask Reliability

We need to think not only about consistency between raters' scores but also about consistency among various tasks we could assign to measure the same

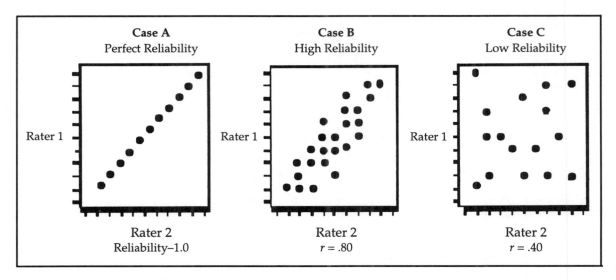

Figure 7.2. Interrater Reliability: A Comparison

knowledge or skill. This issue, called *intertask reliability*, reflects how consistent a student's performance is across certain questions or tasks.

To understand this notion, let's consider a multiple-choice test like the one you took for your driver's license. The California Department of Motor Vehicles asks 18 multiple-choice questions. If they dropped one question and substituted another, do you think you would get the same score? In other words, how consistently does each question measure your knowledge of the "rules of the road." Each question is only one of a much larger pool of questions you could be asked to measure the same knowledge. The DMV expects to be able to generalize, from one sample of questions, how well you could answer all the conceivable questions they could ask.

With a performance assessment, students have at best a few tasks instead of many individual questions. Consider the House task you did in Chapter 2. You were asked to construct circuits to create two signals to tell a friend when to visit (i.e., two tasks). These tasks were designed to measure your knowledge of electric circuits, and we could have posed other tasks to assess the same knowledge.

Intertask reliability, then, asks how interchangeable performance is from one task to other tasks you could have used to measure performance.

We examine intertask reliability the same way we examine interrater reliability, by a reliability coefficient (0 to 1), reflecting the correlation between two or more tasks. Figure 7.3 shows what intertask reliability coefficients look like on a scatter plot for perfect, high, and low reliability. Here we are looking at three tasks for a group of four students. With perfect reliability (1.0) the students are rank ordered the same for each task. With a reliability coefficient of .80, we see that although the students who were highest (and lowest) switched places on one or another of the tasks, they still remained essentially in the same order. No one who was very high on one task scored very low on another. This is not the case with low intertask reliability (Case "C"). Notice

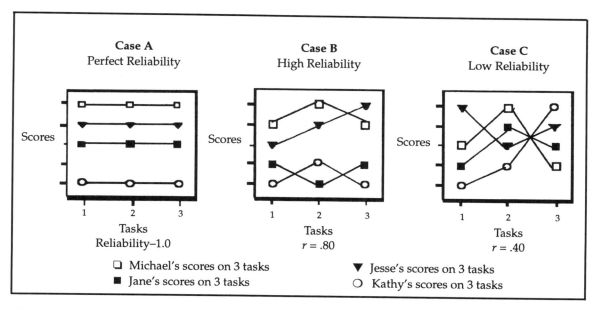

Figure 7.3. Intertask Reliability: A Comparison

that the student who was lowest on the first task was highest on the third task—and the rest were just as diverse. Obviously, performance across the tasks was not consistent.

Our research has indicated that, typically, intertask reliability is low to moderate for performance assessments with only a few tasks. This tells us that any one task might not be representative of how a student would do on another task. In choosing performance assessments, therefore, you should ensure that the students are provided enough tasks to allow them ample opportunity to demonstrate what they know and can do.

Validity

Not only must an assessment be reliable, it must also be valid. Validity asks the question, "Is the assessment tool measuring what it is supposed to be measuring?" There are two kinds of validity that teachers need to be particularly aware of: *content validity* and *exchangeability*.

Content Validity

If you are going to use a particular performance assessment to monitor the effectiveness of learning activities in your classroom, there needs to be a match between what is taught and what is tested. This point seems all too obvious, but it can be tricky in practice. Unlike reliability, there is no numerical coefficient to use for content validity. It relies on the professional judgment of the teacher regarding whether students have had an opportunity to learn the concepts and procedures being tested in the performance assessment.

Here are some important questions to ask yourself when considering the content validity of a performance assessment for a particular science unit:

1. Does there appear to be a parallel between the instructional goals of the lesson/unit and those that the performance assessment is measuring?
2. Are the ways students are asked to respond in the assessment similar to the ways students respond in classroom activities?

Our research team has found that performance assessments tend to be content valid if they are created specifically for a hands-on science curriculum and if the tasks in the assessment are taken from the activities in the unit or if the tasks extend activities in it.

Let your professional judgment tell you if a particular assessment is content valid for your purpose: to assess whether students have learned what they have had an opportunity to learn.

Exchangeability of Assessment Methods

Tom, a sixth-grade science teacher, asked us recently, "When I test my students, why can't I just present them with a written problem and have them respond to short-answer questions about that problem? That's the testing method I have always used in the past, and it is a lot less messy than the hands-on investigations you're recommending."

Our research team asked the same question. We wondered whether the scores on paper-and-pencil assessments are exchangeable for those of hands-on performance assessments. And while we're at it, what about computer simulations of hands-on performance assessments? Are these scores exchangeable for those the student would get if he or she could be evaluated by a teacher individually while doing an investigation?

Research Findings on Exchangeability

We examined this issue, using three different performance assessments: Paper Towels, which you did in Chapter 4; Electric Mysteries, which asked students to construct circuits to determine the contents of six black "mystery" boxes; and Bugs, which asked students to determine whether sow bugs choose light or dark, damp or dry, or some combination of these environments. For each assessment, we developed different methods to assess knowledge, keeping the tasks for each assessment as similar as possible. The five assessment methods were these:

- Direct observation
- Notebooks
- Computer simulation
- Short answer (pencil and paper)
- Multiple choice (pencil and paper)

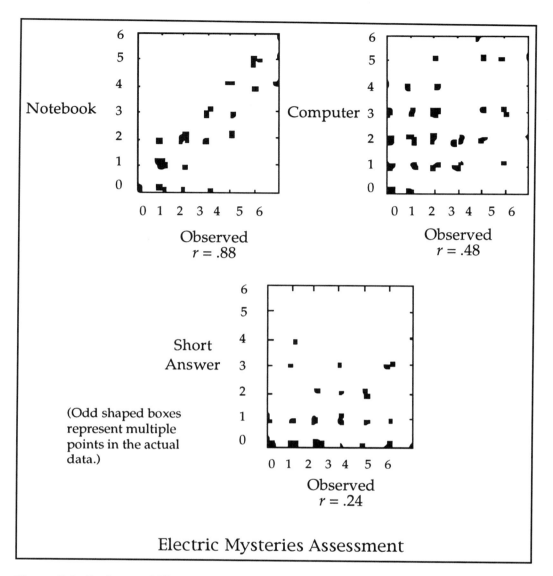

Figure 7.4. Exchangeability

We used the direct observation scores as our benchmark or standard, because we assumed that there is nothing better than working one-on-one with students to understand their level of performance. Then we tested the students on each of the other methods to see how their scores matched with those of direct observation. We found that only the student notebook scores were exchangeable (Baxter & Shavelson, 1994). As you can see from the plots in Figure 7.4, the correlation between direct observation and notebooks is a respectable .88, but the other correlations are considerably lower: .48 for computer simulations and .24 for short answer.

Were you surprised that the computer simulation was not more highly correlated with the hands-on investigations? We were—and a little disappointed, too. It appears that computers measure somewhat different aspects of knowledge than do hands-on investigations, but we're not sure just why this is true. Further research will have to be done to unravel this mystery.

Perhaps you were even more surprised to see how low the correlation is between the short-answer paper-and-pencil test and direct observation. It appears that the nature of the test is changed considerably when students have concrete materials that react to their manipulation—so much so that the two methods are assessing different aspects of knowledge. Perhaps an example will help to clarify this finding.

An Example of Exchangeability

Recall that in the Electric Mysteries assessment, students are asked to figure out what is in a set of black boxes—a battery, a battery and bulb, a wire, two batteries, or nothing at all—and then give the *reasons* for their choices. We found that some students who got perfect scores (6 of 6 correct) on the hands-on investigation did very poorly on the paper-and-pencil test.

Looking at the reasons they gave as evidence for the answers they chose gives us insight into one big problem with paper-and-pencil tests; students gave reasons that they clearly would not have come up with if they had been able to try out their ideas on actual boxes. For example, one student said, "Light come from one box and not the other." A logical answer—but with the real boxes in hand, the student would see that no light was coming from any of the boxes. Another student wrote, "Shake it up if it [made] alot of noise it was a bulb if it didn't it was a wire." Again, a perfectly good answer, except that none of the boxes made noise or even "clunked" when they were shaken. We purposely made the boxes identical, of equal weight, shake proof, and sealed shut so no light could escape. All the students had to do was try out their ideas and they'd know they wouldn't work. The hands-on investigations allowed them to do this, but the paper-and-pencil tests did not.

Notebooks—Good News!

The good news for teachers is that we found that assessment notebooks, like the ones you used when you did House, Paper Towels, Mystery Powders, and Leaves, appear to be exchangeable for direct observation of investigations. Your students can record what they did, and you can score their papers at a later time. The logistics are simplified tremendously.

So the bottom line on this exchangeability issue is that if you want a method of performance assessment that gives essentially the same information you would get if you could look over the shoulders of your students as they proceed through an investigation, choose performance assessments that allow students to manipulate materials and then record their findings in notebooks.

Reliability and Validity Revisited

What is the difference between issues of reliability and those of validity? Think of your friendly bathroom scale as a measurement tool. You get on it one morning and you weigh 130 pounds, and then, 5 minutes later, you get on it again and you weigh 132 pounds; perplexed, you weigh yourself again and this time you weigh 129 pounds. Obviously, you know you have a problem with the *reliability* of your scale. There is *measurement error* that you can't account for. A measurement instrument needs to give consistent (reliable) results.

Now, if you got on the scale to measure your height (this is really stretching it), you are misusing the measurement tool; obviously, it is not measuring what you think it is measuring. This is a *validity* issue: The measurement tool is not valid for the purpose for which you are using it. But let's take this example a step further. Say you visit your doctor on the same day and on her scale you weigh 142 pounds. This is also a validity issue. You are faced with the question of which scale is giving you a measure of your true weight—the doctor's scale or your bathroom scale?

Measurement tools that measure physical characteristics, like height and weight, are a lot easier to interpret than psychological tests that measure knowledge and skills. But all too often, results of tests given in schools are misinterpreted by students and teachers. For example, a test that is thought to measure science knowledge might really be measuring mostly reading and following directions—plus knowledge of some science vocabulary. A well-designed performance assessment offers a way to test students' ability to use the knowledge they have gained to show they can do science—and has the potential for being a valid test of science performance.

When choosing a performance assessment, therefore, reliability and validity are basic to the trustworthiness of an assessment. But we're sure you will agree that two other qualities are of utmost importance: utility and practicality.

Utility

Remember when you were scoring the Mystery Powders assessment in Chapter 5? You were able to use the score form to discover specific weakness in your students' ability to use disconfirming evidence. You could then go back and give students additional learning experiences to help them understand this difficult concept. A scoring system needs to be useful to you, to provide you with clear and specific information about students' conceptual and procedural strengths and weaknesses. When you used the Leaves rubric, you had a more holistic score that could be used for rank ordering students according to level but was not specifically useful for teaching, because you didn't know exactly where each student's problems lay.

Here are some questions to ask yourself about the utility of a performance assessment you are considering using in your classroom:

1. Do the scores provide information helpful in evaluating performance of a student or a class of students?
2. Do the scores provide information helpful in monitoring instruction so that appropriate revisions might be made?
3. Do the scores provide information helpful to the students in judging the quality of their own performance?

Practicality

From talking with teachers, we know that this issue hits closest to home. If an assessment is not practical—easy to administer and cost efficient in terms of both time and money—then it is not likely to be used.

We have found, however, that a teacher's background in teaching hands-on science is a tremendous predictor of his or her acceptance of performance assessments. Those teachers who are not new to hands-on science have developed the logistics of passing out materials, maintaining clutter control, and dealing with other classroom management issues necessary to handling most performance assessments with ease. After all, a performance assessment looks and feels like just another learning activity—with a score form.

When judging a performance assessment in terms of ease of administration, compare it with hands-on learning activities—not multiple-choice tests. The two are simply not comparable; they do not measure the same knowledge. That is not to say, however, that all performance assessments are equal in practicality. Only you can be the judge of how doable performance assessments are in your classroom.

Cost is also an important factor. Materials that are reusable may require an outlay of money the first year but will be useful for years to come. If equipment is already available in your hands-on science kits, so much the better; however, monetary cost is definitely a factor to consider.

Another consideration is cost in time spent—both your time and that of your students. Many teachers are reluctant to spend one to two class periods testing students, because they think learning stops while testing goes on. Because performance assessments are learning activities, however, teachers have found that time is really not wasted. Students do continue learning while doing the assessments and, in fact, might make important connections while thinking through the task.

The time you need to spend scoring is also a consideration. We suggest you examine the score form when evaluating a performance assessment. It should be clear and easy to use with a little practice. By the time you score 10 students, you should be up to speed and taking no more than 1 to 2 minutes to score each student.

Here are some questions to ask yourself when looking at the practicality of a performance assessment:

1. Is the assessment manageable?
2. What organizational steps need to be taken to make the assessment manageable?
3. How is the task performed (individually or by pairs)?
4. How long does it take to set up the task?
5. How long does it take to administer the task?
6. How long does it take to score each student (after having practiced scoring about 10 students)?
7. How much does the performance assessment cost per student?

Summary

Choosing a performance assessment that you can trust to give you reliable and valid information about what students know and can do in science is vital to you and your students. Let's face it: Tests are powerful. They also need to be useful and practical or they are more trouble than they're worth. In selecting a performance assessment to use in your classroom, you need to keep four qualities in mind: reliability, validity, utility, and practicality.

Assessments need to give consistent results no matter who is scoring the assessment (interrater reliability) or which particular task students are asked to perform to test their knowledge (intertask reliability). Teachers should request the interrater reliability coefficients for assessments and check to see if they are above .80. If intertask reliability coefficients are low or moderate, teachers should not rely too heavily on any one task but should give students several opportunities to show what they can do.

Perhaps the most important quality a performance assessment should have is that it must overlap the curriculum for which it is used; it must be content valid. One way to check this match is to ask yourself if the assessment would make a good learning activity for the unit. It should look and feel like the next logical step in the learning sequence or the culminating activity in the unit. Teachers use their professional judgment to determine the content validity of a particular assessment, asking themselves if the content and format of the assessment test what the students have had an opportunity to learn.

Our research has shown that scores from assessment notebooks are highly correlated with those from direct observation, allowing us to have confidence that we can use these methods interchangeably. Unfortunately, computer simulations and short-answer and multiple-choice pencil-and-paper tests are not as exchangeable for direct observation and appear to measure somewhat different aspects of knowledge.

Issues of usefulness and practicality are very important to harried and overworked teachers. Assessments must provide useful information about the specific concepts and procedural knowledge students either have or are still weak on, and they should be cost effective in terms of time and money.

This chapter has contained a lot of "heavy" stuff, but now we'll pull it all together. In the next chapter, you will be asked to apply what you have learned by choosing the best among four possible assessments for a unit on electricity.

Which Performance Assessment
Is Best for Your Purpose?

This chapter gives you a chance to apply the principles you've learned throughout this book. It invites you to put yourself in the place of a teacher who is planning to give a performance assessment at the end of a unit on electricity. You will examine four possible performance assessments, weigh the pros and cons of each, and then come to a decision about which is best for your needs.

First, we present you with an overview of an electricity unit so you can get a feel for the goals of the unit through the various learning activities the students experience. This will help you judge the content validity of the assessments. Then you will carefully examine each assessment itself, noting how each satisfies the qualities of trustworthiness you learned about in Chapter 7.

We conclude this chapter by letting you listen in on discussions we led with groups of teachers, debating the pros and cons of the four assessments you examined. See if your conclusions match theirs.

Choosing a Performance Assessment: An Exercise for You to Try

Choosing the right performance assessment for your class, one that is both a trustworthy measurement tool and consistent with the goals of your unit, is no mean trick. You might find yourself weighing the advantages and disadvantages of several possible performance assessments, or, more likely, you might find a suggested performance assessment every so often throughout the teacher's manual of your science curriculum and wonder how it stacks up—is it worth giving at all?

In this chapter, we provide an exercise that will allow you to evaluate four different assessments that could be used at the end of a unit on electricity. This selection exercise includes the following:

1. An overview of Circuits and Pathways, a fifth- /sixth-grade unit on electricity (Figure 8.1)

2. A chart providing criteria for judging assessments (Figure 8.2), together with an additional page for notes

3. Four performance assessments, A, B, C, and D, for you to analyze and evaluate (Figures 8.3, 8.4, 8.5, and 8.6)

Finally, there is a discussion of how other teachers have evaluated these assessments in terms of the criteria presented in Figure 8.2.

Overview of Circuits and Pathways

Circuits and Pathways[a] is a unit that helps students develop a basic understanding of electricity by exploring simple circuits made with batteries, wires, bulbs, and motors. The purpose of the unit is to help students understand that

1. batteries, wires, motors, and bulbs can be connected in specific ways to create electrical circuits;
2. electrical current has direction (it matters which way the battery is facing in the circuit); and
3. by manipulating circuit components, you can change bulb brightness, motor speed, and motor direction.

Overview of the Unit

- Exploring and recording various ways of constructing circuits with motors and batteries to make the motors spin in different directions (polarity)
- Learning to distinguish complete from incomplete circuits first by causing a motor to spin and then by causing a bulb to light
- Learning to draw complete circuits
- Assembling different types of complete circuits to light a bulb (series and parallel circuits)
- Learning to classify and compare materials that are conductors and nonconductors of electric currents
- Using standardized techniques to measure the relative brightness of bulbs in various circuit configurations
- Learning to build a switch

a. Unit developed by the Educational Development Center. Available from Optical Data School Media (see Resource D for address).

Figure 8.1. Overview of Circuits and Pathways

Criteria for Judging Assessments	A	B	C	D
Does the assessment meet the requirements for being a performance assessment: task, response format, and scoring system?				
Does there appear to be a parellel between instructional goals of the unit and that which the assessment is measuring? **(content validity)**				
Is a value available for **interrater reliability**? If so, what is it?				
Is a value available for **intertask reliability**? If so what is it?				
Does the assessment appear to be exchangeable for direct observation? **(exchangeability)**				
Is the information from scoring the assessment useful for: (1) curriculum monitoring and/or (2) accountability **(utility)**				
Would use be reasonable in terms of: (1) impact on the classroom, (2) teacher time required, and (3) cost? **(practicality)**				

Figure 8.2. Criteria for Judging Assessments (1 of 2)

ADDITIONAL NOTES

Assessment A

Assessment B

Assessment C

Assessment D

Figure 8.2. Criteria for Judging Assessments (2 of 2)

PERFORMANCE ASSESSMENT A
Unit on Electrical Energy

The Task:

Pretend you are helping your younger sister build a camper-van for her Barbie dolls out of a cardboard box. She wonders if it would be possible to have two lights in the van that really work. You have been studying electric circuits in school, so you offer to give it a try. Imagine that you have 2 batteries, 2 bulbs, and 6 pieces of wire to work with, along with electricians' tape—and a lot of creativity!

1. Draw a diagram of the circuit you would construct in the van.

2. How are the lights turned on and off?

Figure 8.3. Performance Assessment A (1 of 1)

PERFORMANCE ASSESSMENT B
Electric Mysteries

Summary of the Task	Students are asked to find out what is inside six mystery boxes. For each box students have to connect a circuit to help themselves figure out what is inside each box.
Curriculum Topic	Electricity
Grade Level	Fifth-Sixth Grade
Prerequisite Knowledge	Students should have some background knowledge about assembling electric circuits, polarity, and brightness of bulbs in various circuit configurations.
Equipment	Six high quality metal boxes, two batteries, two bulbs, five wires. $15 per set-up
Technical Characteristics	Perfect reliability between raters, moderate intertask reliability and moderate stability have been obtained for sixth grade students. Correlation with multiple-choice science achievement tests is .31.
Time of Administration	30 to 60 minutes
Prior Preparation	The teacher needs to set up a table with the equipment before the task is performed. Equipment has to be checked before each administration.
Instructions	The teacher should read the instructions for the task aloud, introduce the equipment to the students, and encourage them to write the answers in the notebook.
Scoring	To get a point for each mystery box, the student has to correctly identify the contents and draw the correct circuit. The score ranges from 1 to 6.

Figure 8.4. Performance Assessment B (1 of 8)

ELECTRIC MYSTERIES

Name _____

You have some batteries, bulbs and wires in front of you for doing some experiments. All the wires are the same. They are just different colors. They have clips on the end so you can connect things.

PART I

1.   **Connect one battery, one bulb and wires so the bulb lights.**

Draw a picture in the box that shows what you did. If you like you can use symbols like these:

Battery symbol: —| |—

Bulb symbol:

Wire: _____

Figure 8.4. Performance Assessment B (2 of 8)

2.   **Figure out what is in the mystery box labeled with a question mark "?".**

The box has inside it either a battery or a wire:

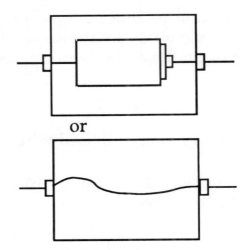

To help figure out which one is in it, connect it in a circuit with a bulb:

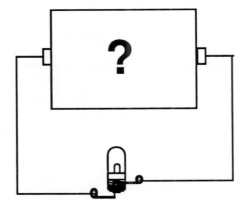

Fill in the answer:

The "?" Box has a _____ in it.

Figure 8.4. Performance Assessment B (3 of 8)

PART II

Find out what is in the six mystery boxes A, B, C, D, E and F. They have five different things inside, shown below. Two of the boxes will have the same thing. All of the others will have something different inside.

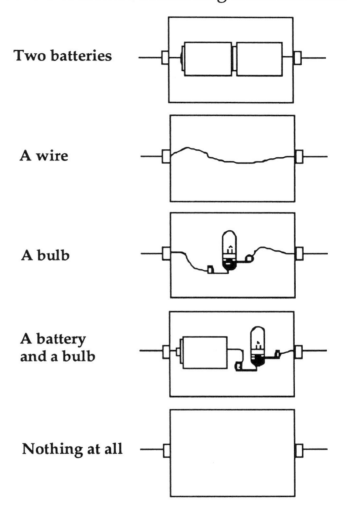

Two batteries

A wire

A bulb

A battery
and a bulb

Nothing at all

For each box, connect it in a circuit to help you figure out what is inside. You can use your bulbs, batteries and wires any way you like.

When you find out what is in a box, fill in the spaces on the following pages.

Figure 8.4. Performance Assessment B (4 of 8)

Name: _____

Box A: Has _____ inside.

Draw a picture of the circuit that told you what was inside **Box A**

How could you tell from your circuit what was inside **Box A**?

_____

_____

Box B: Has _____ inside.

Draw a picture of the circuit that told you what was inside **Box B**

How could you tell from your circuit what was inside **Box B**?

_____

_____

Figure 8.4. Performance Assessment B (5 of 8)

Box C: Has _____ inside.

Draw a picture of the circuit that told you what was inside **Box C**

How could you tell from your circuit what was inside **Box C**?

_____

_____

Box D: Has _____ inside.

Draw a picture of the circuit that told you what was inside **Box D**

How could you tell from your circuit what was inside **Box D**?

_____

_____

Figure 8.4. Performance Assessment B (6 of 8)

Box E: Has _____ inside.

Draw a picture of the circuit that told you what was inside **Box E**

How could you tell from your circuit what was inside **Box E**?

_____

_____

Box F: Has _____ inside.

Draw a picture of the circuit that told you what was inside **Box F**

How could you tell from your circuit what was inside **Box F**?

_____

_____

Figure 8.4. Performance Assessment B (7 of 8)

Electric Mysteries End-of-Unit Assessment Score Form

Student _____   Scorer _____

Mystery Box	What's Inside	Circuit	
A	a battery and bulb ⬭	A	▢
B	a wire ⬭	B	▢
C	nothing ⬭	C	▢
D	two batteries ⬭	D	▢
E	a bulb ⬭	E	▢
F	a wire ⬭	F	▢
			Total ⬜/6

Figure 8.4. Performance Assessment B (8 of 8)

PERFORMANCE ASSESSMENT C
Unit on Electrical Energy

Materials for each setup:

1. 2 D batteries

2. 2 battery holders

3. 2 bulbs

4. 2 bulb holders

5. 6 pieces of wire

Series and Parallel Circuits

1.   Build each of the following circuits. Then, on the next page, record on the chart what happens when you:

   a)   remove one bulb from its socket in the circuit
   b)   remove one battery from its holder in the circuit

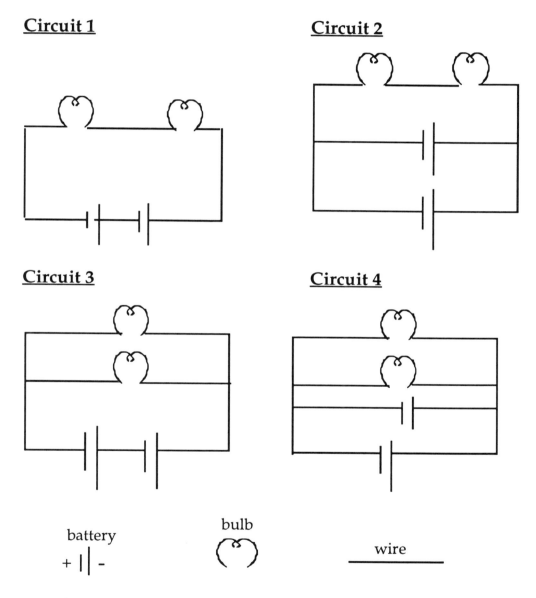

Circuit 1

Circuit 2

Circuit 3

Circuit 4

battery

+ || -

bulb

wire

Figure 8.5. Performance Assessment C (1 of 3)

WHAT HAPPENS WHEN I:

Circuit	remove one bulb from the socket	remove one battery from the holder
1		
2		
3		
4		

2. On the back of this paper show how many different complete circuits you could construct using 3 batteries and 3 bulbs (with no more than 8 pieces of wire). Draw a picture of each circuit.

3. Predict what would happen if you removed one bulb and one battery from each circuit just as you did in the first part of this investigation. Construct a chart to show your results.

Figure 8.5. Performance Assessment C (2 of 3)

Answers:

Circuit	remove one bulb from the socket	remove one battery from the holder
1	*light goes out*	*light goes out*
2	*light goes out*	*light stays on*
3	*light stays on*	*light goes out*
4	*light stays on*	*light stays on*

Figure 8.5. Performance Assessment C (3 of 3)

PERFORMANCE ASSESSMENT D
Unit on Electrical Energy

Science: Electricity

Grade Levels: 5-6

Estimated Time: 15 minutes

Materials Provided ($2.50 per station):

Bag 1

Battery—size D
Battery holder with clips
Flashlight bulb
Bulb holder
2 wires

Bag 2

1 wire
Steel paper clip
Steel key
Large (steel) nail
plastic ruler
Wooden toothpick
Cardboard

Instructions:

Set up workstations with bags *1* and *2*.

Assess students either individually or in pairs.

Score, using the holistic scoring rubric provided.

Figure 8.6. Performance Assessment D (1 of 5)

The Task

Directions: In bag *1* you have a battery, bulb, and wires in front of you for
doing some experiments.

1. Take all the materials out of bag *1*.

2. Make a circuit with the battery, bulb, and wires so the bulb lights.

3. Draw a picture of the circuit you made in the box below.

4. Take all the materials out of bag 2.

5. Use the wire to make an electrical conductivity tester, as shown in the
picture below.

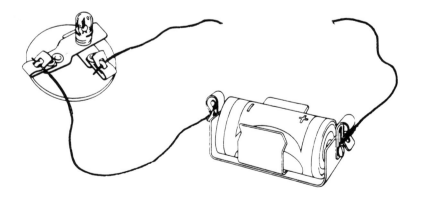

Figure 8.6. Performance Assessment D (2 of 5)

6. Use your conductivity tester to see if the paper clip in bag 2 conducts electricity. Touch the ends of the wires to the ends of the paper clip, as shown in the picture below.

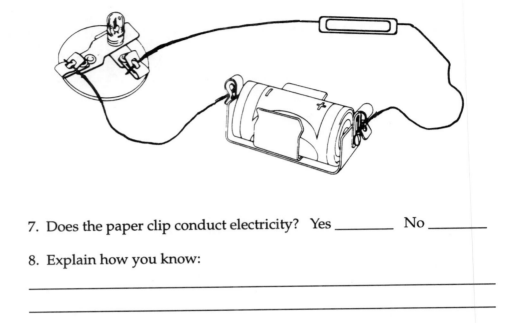

7. Does the paper clip conduct electricity?   Yes _____   No _____

8. Explain how you know:

_____

_____

_____

9. Test the objects in bag 2 with your conductivity tester. Make a chart below to show what happened when each item was tested.

Figure 8.6. Performance Assessment D (3 of 5)

Now answer the questions below.

10. Compare the objects that *do* conduct electricity. How are they alike?

_____

_____

_____

11. Compare the objects that *do not* conduct electricity. How are they alike?

_____

_____

_____

Figure 8.6. Performance Assessment D (4 of 5)

PERFORMANCE ASSESSMENT D
Teacher's Scoring Rubric

Outstanding: Rating = 4

All questions are answered completely and acceptably; gives logical explanations on questions 8, 10, and 11; draws a complete diagram of a circuit; constructs a table that clearly distinguishes conductors from nonconductors; shows understanding of electricity and conductivity.

Competent: Rating = 3

Most questions are answered completely and acceptably; explanations on questions 8, 10, and 11 may be unclear; draws an acceptable current; table may be less sophisticated (i.e., list format); shows understanding of electricity and conductivity.

Satisfactory: Rating = 2

Many incomplete or unacceptable responses; explanation on questions 8, 10, and 11 is very limited; diagram of circuit may be incomplete or missing; shows some understanding of electricity but not conductivity; attempts to answer questions on both pages.

Serious Flaws: Rating = 1

Response severely limited (only diagram or few answers); incomplete answers with no explanation or rationale; no evidence of understanding of the concept of electricity or conductivity; at least one correct answer other than the diagram.

No Attempt: Rating = 0

Figure 8.6. Teacher's Scoring Rubric for Performance Assessment D (5 of 5)

Your Choice

Which assessment do you think would give you the most informative, trustworthy, and otherwise interesting view of your students' conceptual and procedural knowledge in electricity? Was it an easy decision, or are you still pondering the advantages and disadvantages among the various assessments?

Take a few minutes now to look back at your notes, gather your thoughts, and come to a decision. Which assessment would you use and why?

_____

_____

_____

_____

_____

_____

_____

Other Teachers' Choices

You might be interested in how other teachers have viewed these four assessments when faced with the same exercise you just did. Discussed below are the points raised during structured discussions we led with groups of teachers.

Is It a Performance Assessment?

From the outset, the teachers agreed that the first question you have to ask yourself when contemplating an assessment is, "Is this really a performance assessment?" Does it have a *task* (including the manipulation of concrete materials), a *response format*, and a *scoring system*? They agreed that if it does not have these elements, you are probably looking at another learning activity that has not yet been translated into a performance assessment. One of the choices falls into this category: Assessment *A* is an interesting task but includes no scoring system. Also, *A* is does not include the manipulation of concrete materials. It is a paper-and-pencil task done entirely in the abstract. In other words, it is not a true performance assessment.

Content Validity

We then asked, "What about the content validity of the assessments?" The teachers thought that assessments B and D looked pretty good: They appeared to parallel the content of the unit and they would both make useful learning activities for the unit (a good way to check on content validity). D, however, does ask students to construct a table. Because it could be difficult to separate their knowledge of electricity from their ability to organize and present their findings in a table, it is important for the teacher to consider whether the students had the opportunity to learn and practice this procedural skill during the course of the unit. For an assessment to be content valid, both content and procedural skills evaluated in the assessment must overlap the curriculum.

Several teachers noticed that Assessment C has a problem in that it not only gives careful directions for constructing the circuits but also provides a table to be filled out. The students really are only being asked to demonstrate whether they can follow directions, observe what happens, and fill in a chart. The students are then asked to infer what would happen if more batteries and bulbs were added and removed, but no materials are provided to test out the conclusions. It was obvious to the teachers that the unit encouraged more actual learning on the part of students than is captured by this assessment.

Interrater Reliability

Next, the teachers considered the interrater reliability of the four assessments. Because A does not have a scoring system, they obviously could not know whether raters could score it reliably. Assessment B, on the other hand, has perfect reliability. Some teachers questioned how this could be true for a performance assessment, but then they noticed that almost no judgment is required of the teacher. The scoring is very straightforward: The student either gets 1 point for figuring out what is in the box and drawing the one possible circuit that tells him or her so, or gets a 0.

Although no interrater reliability coefficients were available for Assessment C, the teachers could tell from the score form that the first part required no teacher judgment. The cells of the chart were just scored right/wrong, so interrater reliability is not a problem. Questions 2 and 3, however, could lead to unreliability in scoring, because just how they would be scored is not specified.

Assessment D has no reliability information available. Well-defined rubrics have been shown to be reliable, but we don't know about this one.

"So, what do you do if you find an assessment you want to use, but interrater reliability is not provided or it's too low?" one teacher asked. We suggested that if it's not provided, you ask the company or institution that developed the performance assessment to give you reliability information. Otherwise, either you shouldn't use it or you should do an interrater reliability check of your own. You could have pairs of teachers score sets of papers to see whether they rank order students similarly. If it looks like teachers disagree, do not use that assessment.

Intertask Reliability

Some teachers were perplexed about how to judge the intertask reliability of assessments. Only Assessment *B* included intertask reliability information—it's "moderate"—and we've found this situation to be all too common. What should you do about this problem if you want to use one of the other assessments anyway? We suggest you go back to the developer of the assessment and request this information. Otherwise, you could assume that intertask reliability is low to moderate and give students more opportunities to show what they can do, thereby increasing intertask reliability. Go ahead and use this assessment as one piece of information, but add it to other sources of information about what students know and can do.

Exchangeability

The teachers agreed that Assessments *B, C,* and *D* use assessment notebooks, which, research has shown, are exchangeable for direct observation. We can count on them giving us almost the same picture of student performance as if we were standing there watching that performance. Assessment *A* is a paper-and-pencil task that measures somewhat different aspects of knowledge from those gained by direct observation of a hands-on investigation.

Utility

The teachers debated the usefulness of the assessments' scores for monitoring effectiveness of the curriculum and for accountability purposes. Here are their conclusions:

- Obviously, because Assessment *A* has no scoring form, we can't evaluate usefulness of its "scores."
- Assessment *B*, on the other hand, has an analytic score form that allows the teacher to score each student for both result (student's conclusion about what is in the box) and evidence for that result (the circuit that helped him or her reach the conclusion) on the six mystery boxes. This makes it useful for both purposes.
- Assessment *C* has only a right/wrong score form, which is not much help in curriculum monitoring but can give accountability information about the students' ability to follow directions, observe phenomena, and record observations. Remember, however, that this is a highly structured task, giving step-by-step instructions, so information on student knowledge is limited.
- Assessment *D* is scored with a rubric, which is excellent for giving an overall picture of student performance but is not very helpful for curriculum monitoring because it is very general. However, given the fact that this is an end-of-unit assessment, accountability might be all you are looking for.

Practicality

What about the practicality of these assessments? Teachers responded that each of the assessments had its advantages and disadvantages when it came to this important topic.

Time spent scoring *A* is unknown, because there is no scoring system. It is very easy to administer, as are all paper-and-pencil tests. This ease of administration, however, must be viewed in light of its limited usefulness as a performance measure, because it is only a paper-and-pencil test.

B, on the other hand, is expensive, although the materials are reusable and could be shared with other classrooms. *C* and *D* are more moderately priced.

What's the Bottom Line?

Which assessment would you choose for the end of your unit? As you can see, the decision is not straightforward but depends on many considerations, some involving the inherent quality of the assessment and others involving your specific situation and needs.

We hope that this selection exercise has given you some experience in looking critically at a few assessments. Now that you know what to look for, your hands-on experience assessing actual performance will really be the training ground for what you have learned in this book. As you use performance assessments, you will become increasingly aware of their usefulness to you and your class.

Conclusion

We believe that using performance assessment to assess science knowledge and skills at the elementary and middle school level is an idea whose time has come. Students and teachers alike are recognizing in it a fair, rich, challenging way to get at what students have learned and can do in science. But issues of trustworthiness—reliability and validity—of the performance assessments are crucial issues. As a teacher, you have the future of performance testing in your hands. If you choose your performance assessments carefully, being singularly aware of their quality and appropriateness, then performance assessment can become a valuable new testing technique to add to your other measurement tools, giving you a much more complete picture of what your students know and can do in science than if you used paper-and-pencil tests alone.

Resource A:
House for House Assessment

Each assessment setup includes one of these "houses." Cut out along the dotted lines and mount on tagboard for strength. Then cut out the circle below the light and place a bulb (in its bulb holder) through the hole.

Resource A. "House" for House Assessment
Brown and Shavelson, *Assessing Hands-On Science: A Teacher's Guide to Performance Assessment.* Copyright 1996, Corwin Press, Inc.

Resource B:
Student Lab Notebook

During the course of the Mystery Powders unit, students explore the chemical and physical attributes of the five powders, keeping notes in a lab notebook such as the sample notebook found on the next three pages. The notebook (cut in half and stapled together) can be used by the student—as well as by you, the reader—as a reference when taking the assessment.

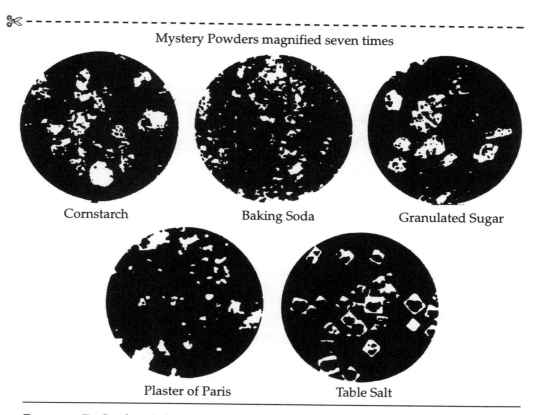

Mystery Powders magnified seven times

Cornstarch   Baking Soda   Granulated Sugar

Plaster of Paris   Table Salt

Resource B. Student Lab Notebook (1 of 3)

TOUCH TEST

Powder Name	What happens with TOUCH
Baking Soda	tiny grains
Cornstarch	slippery
Salt	grainy
Sugar	grainy
Plaster of Paris	sticks to your fingers

✂ -

WATER TEST

Powder Name	What happens with WATER
Baking Soda	dissolves
Cornstarch	turns water white, stays there
Salt	dissolves
Sugar	dissolves
Plaster of Paris	turns water white, sits there, then hardens later

Resource B. Student Lab Notebook (2 of 3)

VINEGAR TEST

Powder Name	What happens with VINEGAR
Baking Soda	fizzes
Cornstarch	doesn't fizz
Salt	disappears
Sugar	disappears
Plaster of Paris	melts, hardens

✂ -

IODINE TEST

Powder Name	What happens with IODINE
Baking Soda	turns yellow
Cornstarch	turns purple
Salt	yellow
Sugar	yellow
Plaster of Paris	yellow - flat

Resource B. Student Lab Notebook (3 of 3)

Resource C:
Alternate Scoring for
Mystery Powders Assessment

Recall that you added a little mystery to the Mystery Powders assessment by randomly choosing two of the four Baggies you had mixed yourself. Here is the score form you would use to score the other two mixtures: baking soda and sugar, and cornstarch and baking soda.

Although our embedded assessment uses baking soda and salt, and plaster of paris and salt, these alternative combinations do show up on the end-of-unit assessment, which asks students to identify the contents of six Baggies containing one or two powders each.

We have also included a copy of the original Mystery Powders #2 Score Form for your convenience.

Scorer _____   Student _____

POWDER(S) (What's inside the bag?)	OBSERVATIONS (How did you know? What happened?)			TEST(S)
	CONFIRMING ↓	DISCONFIRMING ↓	OTHER ↓	

BAKING SODA(D) and SUGAR(B) ☐
- fizzes, bubbles ◯
- turns yellow ◯ — iodine ◯
- vinegar ◯
- dissolves ◯ — water ◯
- grainy ◯ — touch ◯
- sight ◯
- irregular crystles ◯

☐

CORNSTARCH (A) and BAKING SODA(D) ☐
- turns purple,black ◯
- fizzes, bubbles ◯
- doesn't dissolve ◯ — iodine ◯
- vinegar ◯
- water ◯
- not grainy ◯ — touch ◯
- no crystals ◯ — sight ◯

☐

TOTAL Correct Answer ☐ 2

TOTAL Quality of Evidence ☐ 8

QUALITY OF EVIDENCE

4 All **black** AND all ⬜white⬜

3 One **black** AND one or more underlined AND/OR ⬜white⬜

2 One **black** only OR all ⬜white⬜

1 One ⬜white⬜ AND / OR one or more underlined

0 Nothing relevant or tests without observations

NOTE: Subtract 1/2 point if student records 1 or more observations without a corresponding test. Maximum deduction of 1/2 point per Mystery Powder

---

Resource C. Mystery Powders Alternate Score Form
Brown and Shavelson, *Assessing Hands-On Science: A Teacher's Guide to Performance Assessment.* Copyright 1996, Corwin Press, Inc. Alternate Scoring Form

Scorer_____   Student_____

POWDER(S) (What's inside the bag?)	OBSERVATIONS (How did you know? What happened?)			TEST(S)	
	CONFIRMING ↓	DISCONFIRMING ↓	OTHER ↓		
X BAKING SODA (D) and SALT (C) ☐	fizzes, bubbles ⬭ cube-shaped crystals ⬭		turns yellow ◯- dissolves ◯- grainy ◯-	iodine ◯ vinegar ◯ water ◯ touch ◯ sight ◯	☐
Y PLASTER (E) and SALT (C) ☐	cube-shaped crystals ⬭	turns yellow not black ⬭ doesn't fizz ⬭ powdery ⬭	doesn't dissolve ◯- grainy ◯-	iodine ◯ vinegar ◯ water ◯ touch ◯ sight ◯	☐

TOTAL
◨ **2** Correct Answer

TOTAL
Quality of Evidence ◨ **8**

QUALITY OF EVIDENCE

4   All   [black]   AND   all   [white]

3   One   [black]   AND   one or more underlined AND/OR   [white]

2   One   [black]   only OR   all   [white]

1   One   [white]   AND / OR one or more underlined

0   Nothing relevant or tests without observations

NOTE: Subtract 1/2 point if student records 1 or more observations without a
       corresponding test. Maximum deduction of 1/2 point per Mystery Powder.

Resource C. Mystery Powders Embedded Assessment #2 Score Form (Original)

Brown and Shavelson, *Assessing Hands-On Science: A Teacher's Guide to Performance Assessment.* Copyright 1996, Corwin Press, Inc.

Resource D:
Where to Obtain
Instructional Units

To order instructional units, request "Insights Teacher Guide" for "Circuits and Pathways" (House Assessment and Electric Mysteries) and "Mysterious Powder" (formerly called "Mystery Powders").

Optical Data School Media
30 Technology Drive
Warren, NJ 07059
(800) 524-2481

References and Suggested Readings

Baxter, G. P., & Shavelson, R. J. (1994). Science performance assessments: Benchmarks and surrogates. *International Journal of Educational Research, 21*(3), 279-298.

Baxter, G. P., Shavelson, R. J., Goldman, S. R., & Pine, J. (1992). Evaluation of procedure-based scoring for hands-on science assessment. *Journal of Educational Measurement, 29*(1), 1-17.

Brown, J. H., & Shavelson, R. (1994). New ways to measure what students know and can do: A step-by-step guide to using performance assessment for hands-on science. *Instructor, 103*(7), 58-61.

Brown, J. H., & Shavelson, R. J. (1994). Does your testing match your teaching style? *Instructor, 104*(2) 86-89.

Ruiz-Primo, M. A., Baxter, G. P., & Shavelson, R. J. (1993). On the stability of performance assessments. *Journal of Educational Measurement, 30*(1), 41-53.

Shavelson, R. J., & Baxter, G. P. (1992). Linking assessment with instruction. In F. Oser, J. L. Patry, & A. Dick (Eds.), *Effective and responsible teaching: The new synthesis* (pp. 80-90). San Francisco: Jossey-Bass.

Shavelson, R. J., Baxter, G. P., Copeland, W., Ruiz-Primo, M. A., Decker, L. D., Brown, J. H., & Druker, S. (1994). *Performance assessment in science, a teacher enhancement program.* University of California, Santa Barbara. For more information, contact Shavelson Research Laboratory, School of Education, Stanford University.

Shavelson, R. J., Baxter, G. P., & Gao, X. (1993). Sampling variability of performance assessments. *Journal of Educational Measurement, 30*(3), 215-232.

Shavelson, R. J., Baxter, G. P., & Pine, J. (1991). Performance assessments in science. *Applied Measurement in Education, 4*(4), 347-362.

Shavelson, R. J., Baxter, G. P., & Pine, J. (1992). Performance assessments: Political rhetoric and measurement reality. *Educational Researcher, 21*(4), 22-27.

Shavelson, R. J., Carey, N. B., & Webb, N. B. (1990). Indicators of science achievement: Options for a powerful policy instrument. *Phi Delta Kappan, 71*(9), 692-697.

Shavelson, R. J., Copeland, W. D., Baxter, G. P., Decker, D. L., & Ruiz-Primo, M. A. (1994). In S. J. Fitzsimmons & L. C. Kerpelman, *Teacher enhancement for elementary and secondary science and mathematics: Status, issues, and problems* (5/1-5/45). Washington, DC: National Science Foundation.

Tamir, P. (1993). A focus on student assessment. *Journal of Research in Science Teaching, 30*(6), 535-536.

Vavrus, L. (1990). Put portfolios to the test. *Instructor, 100*(1), 48-53.

Watson, B., & Konicek, R. (1990). Teaching for conceptual change: Confronting children's experience. *Phi Delta Kappan, 71*(91), 680-685.

Wiggins, G. (1989). A true test: Toward more authentic and equitable assessment. *Phi Delta Kappan, 70*(9), 703-714.

Wiggins, G. (1992). Creating tests worth taking. *Educational Leadership, 49*(8), 26-33.

Wittrock, M. C. (1991). Cognition and testing. In M. C. Wittrock & E. L. Baker (Eds.), *Testing and cognition* (pp. 1-4). Englewood Cliffs, NJ: Prentice Hall.